SHA'Z SQUARE

Ancient Mysteries Decoded

2nd edition

Shahrokh Zadeh

Ancient Mysteries Decoded ® series

CITI OF
BOOKS

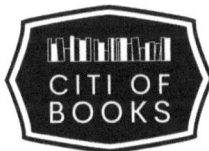

CITIOFBOOKS, INC.
3736 Eubank NE Suite A1
Albuquerque, NM 87111-3579
www.citiofbooks.com

| Hotline: | 1 (877) 389-2759 |
| Fax: | 1 (505) 930-7244 |

Ordering Information:

Quantity sales. Special discounts are available on quantity purchases by corporations, associations, and others. For details, contact the publisher at the address above.

Printed in the United States of America.

| ISBN-13: | Softcover | 979-8-89391-675-1 |
| | eBook | 979-8-89391-676-8 |

Library of Congress Control Number: 2025909615

If you are curious, you'll find the puzzles around you.
If you are determined, you will solve them.

Erno Rubik

I give heartfelt thanks to my family and especially those who have supported me on this journey.

Contents

Introduction

Understanding our past allows us to understand our present. There are many mysteries in our ancestors' culture that we seek to understand. With research, we began to see crumbling monuments and the thought process of an ancient world. As long as 20,000 years ago, evidence on an ancient artifact known as the Ishango Bone existed. The review of antiquity structures and artifacts sheds light on the concerns and interests of a different era.

Imagine our world without numbers. Common sense and ancient evidence point to the idea that numbers and counting began with the number 'one.' Historians believe numbers and counting expanded beyond the number 'one' about 4,000 BCE somewhere in Southern Iraq. Despite its importance, the development of numbers remains mostly a mystery. Galileo Galilei explains how the universe looks like a big book written in the mathematical language. I think it makes complete sense due to the variety of footprints everywhere in nature and even in our life and our communicational methods. We can see simple ones and perplexing ones combined. Mathematics is not just numbers. Geometry is an original field of mathematics. It is the oldest of all sciences, going back at least to Euclid of Alexandria, Pythagoras, and other natural philosophers of ancient Greece. Numbers and geometry are tightly woven together all over the universe. Geometry is concerned with shape, size, the relative position of figures, and space properties. You can see a simple to sophisticated geometry everywhere in nature, such as in a small flower, on our faces, and in our solar system. Of course, our ancestors had seen it as well. They learned how to build magnificent structures and artifacts and learned how to navigate and send messages to each other by geometry. It seems like the ancient engineers were sometimes sending some messages to future people by building magnificent structures in precise shape, size, and direction that many still baffle our scientists.

Certain places around the globe had also baffled scholars for years. There are many things today that we are still trying to understand, such as structures that were very well thought out and intentionally designed and built-in specific locations, with particular directions. People wonder how our ancestors figured out to combine mathematics with spots on the surface of Earth. Many structures appear to have been geometrically crucial in their places, which are related to each other. The ancient engineers demonstrated their use of circles, squares, rectangles and triangles by building the constructions in those specific locations. It now became more comfortable to believe that they chose both shapes and areas of the structures carefully. As David Hilbert once said, "Mathematics knows no races or geographic boundaries; for mathematics, the cultural world is one country."

In this book, I will delve into a few of the old structures globally, such as the Ziggurats in the Middle Eastern area and the great pyramids in Egypt, to represent how they were built based on mathematics and they correlate to each other geometrically by their location and direction.

Each time we discover something new, we find surprises. As Nicolaus Copernicus said, "I am aware that a philosopher's ideas are not subject to the judgment of ordinary persons, because it is his endeavor to seek the truth in all things, to the extent permitted to human reason by God."

The more you read here, the more questions you may have. That is good! History and archaeology, combined with mathematics, are scientific topics that require digging deep. The moment you uncover something, you will find it can lead to something else entirely. So, you might even have more questions than answers that I respectfully refer to the related specialists.

All measurements and figures in this book are exclusively for the reader's reference and from reliable sources such as the Google Earth software supported by Google. You may notice that there are some lines, circles and name tags in some figures. I created them with the use of the software's features for an easier understanding of the content.

This article respects all religions globally, including but not limited to Christianity, Judaism, Buddhism, Hinduism and Islam. Discussion and subsequent judgment of my research are at your (the reader's) discretion. It is not my intent to cause any fear or confusion, but instead to generate knowledge and education.

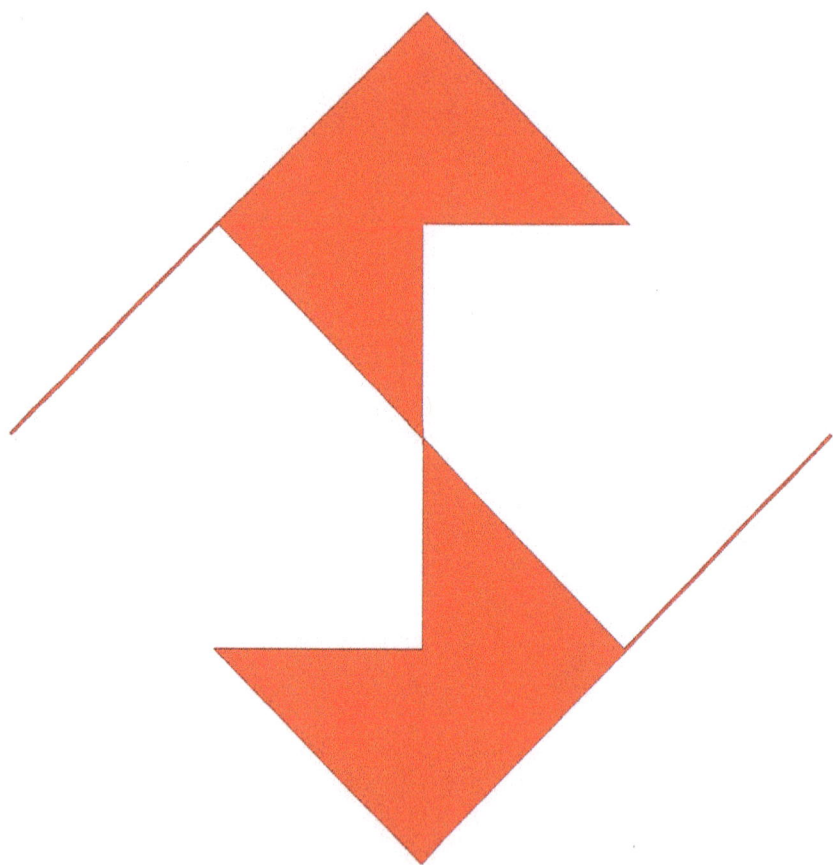

Part 1
Sha'Z Square

This section is a study of six ancient human-made structures that share a typical geometric relationship. I want to begin with a brief explanation of the Modern Geographic Coordinate System.

Modern Geographic Coordinate System

As you might already know, a geographic coordinate system is a threedimensional coordinate system that enables every location on Earth to be specified by a set of numbers, letters or symbols. The invention of the geographic coordinate system was in the third century BCE. It improved a century later to use latitude and longitude, which measures the angles (in degrees) from the center of the earth to a point on the earth's surface where Earth is a sphere model. In 1884, the scientists agreed to adopt the Royal Observatory's longitude in **Greenwich, England,** as the zero reference line.[1][2]

In this article, I added the coordinates of locations for study and comparison. I used the Google Earth program for the measurements.

---

1.   McPhail (2011), the invention of a Geographic Coordinate System is generally credited to Eratosthenes of Cyrene, who composed his nowlost Geography at the Library of Alexandria in the 3rd century BC.

2.   Evans, James (1998), The History and Practice of Ancient Astronomy, Oxford, England: Oxford University Press, pp. 102–103, ISBN 9780199874453.

Ancient Structures

A ***ziggurat*** is a massive structure built in ancient ***Mesopotamia*** and the western ***Iranian Plateau*** (the Persian Plateau). They are usually rectangular-stepped towers, sometimes surmounted by a temple. The structures and their geographic coordinate system used in this part of this article are:

Name	*Location*	*Approximate Built time*	*Coordinate* (Decimal)
The Ziggurat of DurKurigalzu	A- Iraq, near Baghdad	14th century BCE	33.353611°, 44.202222°
The Great Ziggurat of Ur	B- Iraq, near Nasiriyah	21st century BCE	30.962778°, 46.103056°
The Ziggurat of ChoghaZanbil	C- Iran, Kh zest n	1,250 BCE	32.008333°, 48.520833°
The site of Taq-e Bostan	D- Iran, nearKermanshah	4th century AD	34.387529°, 47.132096°
The remains of Persepolis	E- Iran, near Shiraz	515 BCE	29.934444°, 52.891389°
Temple of Apollo, Patroos	F- Greece, Athens	340-320 BCE	37.975542°, 23.722099°

Table 1–Some structures used in this research
(Info: https://en.wikipedia.org and Google Earth)

Ancient Coordination System

Coordinated systems are to specify the position of a point. The structures were in different regions, and it seems there was an awareness of each other's spot. These structures' builders used some coordination system, which appears similar to the modern geographic coordinate cystem we use today.

According to **McPhail (2011)**, Eratosthenes of Cyrene, who composed his now-lost geography at the Library of Alexandria 3rd century BC., invented the geographic coordinate system.[3]

There are six places presented in Table 1, which their latitudes are between 29.93° and 37.98°. These numbers indicate only an 8.05° difference. One thing to consider is that the **Great Ziggurat of Ur**, the **Ziggurat of Chogha Zanbil**, the remains of **Persepolis**, and the **Temple of Apollo** buildings are in the northern hemisphere of the earth. They have a positive latitude number. But what does this mean?

Although these structures are far from the equator and built-in different eras, they have specific latitude numbers based on today's geographic coordinate system units (degrees). Their latitude numbers in degrees (rounded in nearest tenths place) are on the degree units such as 30.0° and 38.0°. Table 2 shows the rounded latitudes of these locations, indicating that four of the six sites are in Group 1, with 0 or 9 in their nearest tenth place, and two sites are in Group 2, with 4 in their nearest tenth place. Their numeral similarities do not seem to be random.

Group	Name	Latitude	Latitude (rounded)
1	The remains of Persepolis	29.934444°	29.9°
	The Ziggurat of Ur	30.962778°	31.0°
	The Ziggurat Chogha Zanbil	32.008333°	32.0°
2	Temple of Apollo, Patroos	37.975542°	38.0°
	The Ziggurat of DurKurigalzu	33.353611°	33.4°
	The site of Taq-e Bostan	34.387529°	34.4°

3.   McPhail, Cameron (2011), Reconstructing Eratosthenes' Map of the World (PDF), Dunedin: University of Otago, pp. 20–24.

Table 2–Some structures used in this research
(Info: https://en.wikipedia.org and Google Earth)

Group 1 consists of 4 structures that, when rounded, have latitudes of 29.9°, 31.0°, 32.0°, and 38.0°.

Group 2 consists of two structures, and both, when rounded, have latitudes ending with 4, with only 1° difference from each other. It seems likely that from 29.9° to 38.0°, some latitudes are missing in their sequence. In other words, there might be some structures correlated with a rounded latitude of 35.0°, 36.0°, and 37.0°.

These facts seem to indicate that our ancestors used a latitude system in their era to build these six structures that, when rounded to their nearest tenth place, is equivalent to 0 or 4 in our modern system. You might want to know if there are other correlations between these six structures.

In the next section, I uncover how geometrically our ancestors built these structures to correlate with each other.

Ancient Location and Geometry

Astronomy has always been an exciting subject for humans, and our ancestors built many structures using this influence to show magnification and other purposes. However, there is reason to believe location and geometry were other vital interests that influenced ancient architecture. We see this reflected in buildings that exhibit a circle, square or rectangle shape.

Notice that the four structures in the following were built in the corners of a large square shape area. There are three ziggurats and the Taq-e Bostan site located in each corner. In this article, I refer to the big, squared area as **Sha'Z Square** for easy distinction. I will discuss each one later in this part. You can further uncover this idea with the use of Google Earth maps. Now, let me review these places, one by one, as follows:

Ziggurat of Dur-Kurigalzu

Figure 1 shows the Ziggurat's side *of Dur-Kurigalzu* (14th Century BC) and its entrance steps from the South view.

Figure 1–Ziggurat of DurKurigalzu and its entrance steps
(https://commons.wikimedia.org/wiki/File:The_Ziggurat_at_ Aqar_Quf.jpg)
Attribution: U.S. Army photo by Spc. David Robbins [CC0]

You can find rectangular and triangular shapes in this instruction, but not any circles or arcs.

Figure 2 shows a Google Earth map of the Ziggurat of Dur-Kurigalzu from above and north wise, which is the large square shape building connected to two straight red lines that I added to the figure for a better understanding of the directions. Some nearby structures underneath the red line are shown in the figure as well, which are square or rectangular shapes with right angles. Modern mapping shows the large square building has about 50° rotation from the north. Note that 50° is 5° more than 45°, which is half the right angle. There are many similar cases in this book that shows small angular differences from a popular geometrical angle. The building is not in a north-south direction. The southeastern red line from the center of the building connects to the Ziggurat of Ur. I explain the northeastern red line, drawn from the square's center later.

Figure 2–Ziggurat of DurKurigalzu
Red lines are added
Google Earth Figure date: 10/9/2010, eye alt: 2,455 ft. north wise.
IMAGE: Maxar Technologies

Ziggurat of Ur

The next figure shows a Google Earth map of the **Ziggurat of Ur** (21st century BCE) and nearby structures from above and northward. The northwestern red line connects to the Ziggurat of Dur-Kurigalzu from the previous section. The building is a rectangular and elongated shape and matches with the northwestern red line. Like the long entrance steps explained in the last paragraph, referencing the Ziggurat of DurKurigalzu, they are also noticeable here and face the northeast. Some nearby structures are shown in the figure as well, which are square or rectangular shapes with right angles. This figure also shows how the ruins in the southern part of the figure has an entrance with a long straight path that is almost parallel with the northwestern red line. This large ruin is practically square.

Figure 3–Ziggurat of Ur Red lines are added
Google Earth Figure date: n 6/21/2010, eye alt: 1,721 ft. north wise
IMAGE: Maxar Technologies

Figure 4–Ziggurat of Ur illustration
https://commons.wikimedia.org/wiki/File:Ziggurat_of_ur.jpg User:
wikiwikiyarou [Public domain]

Note the geometric design of this instruction and compare it with figure 1. Although the Ziggurat of Ur is about seven centuries older than the Ziggurat of Dur-Kurigalzu, its design seems more complicated.

Besides, it appears to look like a giant stepped pyramid. You can find some rectangular and triangular shapes and some arcs, but still, you do not see any circles within the design. Although the builders were familiar with the squared shape structures, they preferred to build the ziggurat in an elongated rectangular shape. You might wonder why. Perhaps the squared building, located at the lower part of figure 3 with a long and straight road in front of it, was meant to be the primary location and facility.

Figure 3 also shows that the long entrance steps are underneath the red northeastern line. The straight and red northeastern line goes directly to the Ziggurat of Chogha Zanbil. Note the Ziggurat of Ur is in the Southern corner of the **Sha'Z Square**, which I will explain later.

Ziggurat of Chogha Zanbil

Figure 5 shows a Google Earth map of the Ziggurat of Chogha Zanbil (1,250 BCE) and its nearby structures from above and north wise. Note that the building is a squared base, with four entrances. There are multiple squares nested in each other in that building. You can also see other structures nearby with square or rectangular bases that have right angles. This prominent squarish building does not include long entrance steps, which is unlike the two previous ziggurats. Note the southwestern red line comes from the Ziggurat of Ur, explained during the last paragraph.

Unlike the Ziggurat of Dur-Kurigalzu, with about 50° turn from the north, this ziggurat has about 45° rotation from the north. Note that 45° is half of a right angle. You might want to know why there is such a geometrical similarity. You might also wonder if the 5° difference in the Ziggurat of Dur-Kurigalzu, was a miscalculation due to lack of precise instruments.

The diameters of the Ziggurat of Chogha Zanbil are north-south and west-east directions. Air mapping shows how if this ziggurat rotates about 20° clockwise, both red lines will match the entrances underneath them. It begs the question of whether the builders made this rotation intentionally.

Figure 5–Ziggurat of Chogha Zanbil
Red lines are added
Google Earth Figure date 1/22/2017, eye alt: 2,552 ft., North view
IMAGE: CNFS/ Airbus

This ziggurat is in the eastern corner of **Sha'Z Square**. The northwest red line goes straight to the Taq-e Bostan, described in the next paragraph.

Figure 6 is a the Ziggurat of Chogha Zanbil taken from the side.

Figure 6–Ziggurat of Chogha Zanbil from side
https://commons.wikimedia.org/wiki/File:Tchogha_Zanbil.jpg Attribution:
Pentocelo [CC BYSA 3.0 (https://creativecommons.org/licenses/bysa/3.0)]

Although our ancestors built this building after the other two structures, you can notice that you do not see many triangles, arcs or circles in its design. The lack of these geometrical shapes makes one wonder if they arranged it deliberately. The reason is unclear.

Taq-e Bostan

Figure 7 shows a Google Earth map of the Taq-e Bostan (4th century A.D.) from above and the north wise. According to Wikipedia, Taq-e Bostan is an artistic site with a series of large rock reliefs, and it means ***Arch of the garden*** or ***Arch made by stone*** in ***Persian, Southern Kurdish***. It is a site from the era of the ***Sassanid Empire of Persia***.

Note on the next figure how the southeastern row of trees and the elongation of the pools are almost parallel with the red lines. The southeastern line goes to the Ziggurat of Chogha Zanbil, while the western red line goes to the Ziggurat of Dur- Kurigalzu.

Figure 7–Taqe Bostan
Red lines are added
Google Earth Figure date 7/6/2019, eye alt: 5,396 ft, North wise
IMAGE: Maxar Technologies

Now, in this section, I want to connect the four discussed places. The illustration below relates the red lines of all four Google Earth maps of the sites showcased in the previous sections. And it presents how the red lines are connected to create a geometrical squared shape (the red lines represent **Sha'Z Square** borders).

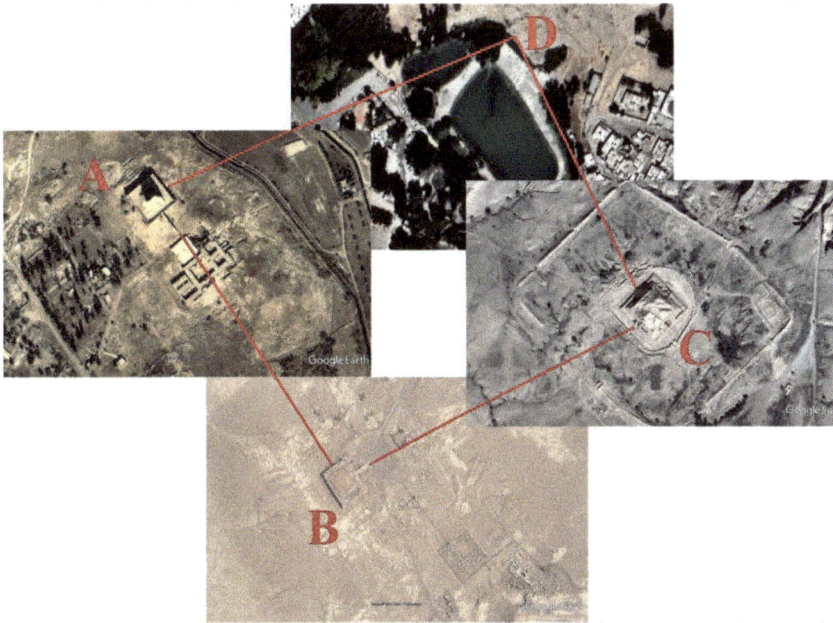

Figure 8–Illustration of Sha'Z Square using previous images
Red Line and letters are added

The illustration is for understanding this matter easier, but the followingfFigure shows how these four places relate in reality from above. The red lines are the borders of **Sha'Z Square**. Note that the red dot in the center of the white circle is not exactly on the diameters' crossline, which means the **Sha'Z Square** is distorted and is not a perfect geometric square shape with four exact right angles.

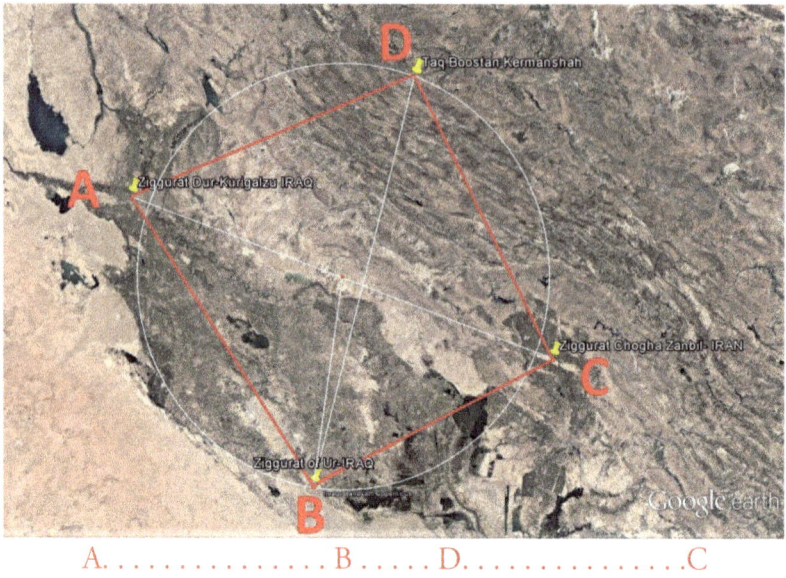

Figure 9–Sha'Z Square area
Red lines are added
Google Earth Figure date 12/2016 eye alt 435.35 mi, North wise
IMAGE: Landsat /Copernicus

The white circle shows how the corners of the **Sha'Z Square** are almost equal distance from each other. Like a diamond or a rhombus shape. Maybe, distortion is due to miscalculations to create the structures in their correct spots, with exact right angles. But what if the engineers designed them in this shape intentionally? No one knows yet.

Note that the geographical coordinate of the intersection spot of the diameters of the **Sha'Z Square** is about 32.632275°, 46.594566°, which I named **CC3** for ease and future references.

If you look at the figure carefully, Spot D is not exactly above Spot B. The corners from left to right are A, B, D, and C. Perhaps there is a reason why there are three ziggurats in each corner of the **Sha'Z Square**, with a king's historical place located in the most northern corner.

Since our ancestors were smart and depicted their thoughts through arts and structures worldwide, I cannot just pass this without searching more. A typical example can be the **Mona Lisa**, a half-length portrait painting by the ***Italian Renaissance*** artist ***Leonardo da Vinci*** in 1503. Leonardo was an artist, an engineer, a scientist, and a mathematician at the same time, so he did not just randomly paint a simple portrait. Each time we study Mona Lisa, we discover surprising messages such as perspectives, the golden ratio, and mathematics.

Maybe the answer to the ***Sha'Z Square*** shape (a rhombus) and its importance is related to the next picture (figure 10), which shows an artistic curve inside the site. It offers one king (red color) looking west and sitting on a cubic chair surrounded by three men looking at him. To me, it resembles the corners of ***Sha'Z Square***; two men on the left representing locations of A and B and one man on the right side (teal color) representing location C, and the king represents location D.

Perhaps the cubic chair is a sign that is in the poses of the king. It might represent glory and wisdom. Also, note how the chair seems empty, and its rhombus side is like the shape of ***Sha'Z Square*** from above.

A. . . . . B . . . . . D. C

Figure 10–A rockface relief from Dowlatshah at Taqe Bostan
Attribution: Sahand Ace [CC BYSA 3.0
(https://creativecommons.org/licenses/bysa/3.0)]

13

Note how the king's arms and hands create a square gesture, while the other three men have different hand gestures.

Also, observe the curve again. You can see a frame with some writings in Farsi between man B and King D's heads.

The top's frame is in the center of the big arc above the men. But why isn't King D placed in the center? Well, by comparing with the **Sha'Z Square** shape and corners, I think the artist might have used his creativity to represent the center of **Sha'Z Square**, which is between the corners of B and D in figure 9. You can easily see a rhombus shape like Sha'Z Square shape in the Farsi frame's lower edge above their head if you look carefully.

You might also find some other interesting things here, such as their swords or the direction of their shoes towards the center of the art, but let us move on to the next section and continue our research.

Persepolis

The Persepolis structures in Iran were in the sixth century B.C. before Taq-e Bostan. There are many pictures of the site from different perspectives, available online, and in the media. By looking at them, you will see many geometric shapes, including circular pillars and sculptures of humans and creatures in Persepolis.

The next figure is a Google Earth map of the area. The green shape, which looks like a rectangle, shows the borders of the building. In the lower part of the figure, you can see some other rectangular structures as well. This kind of design makes it different from the other two ziggurats that I discussed earlier. It makes you think that some evolution in their plans happened there.

Even though the Persian designers knew how to use circles in their structures, they still built their buildings in a rectangular shape. An example would be on the left side of the figure; you can see a prominent modern circular structure almost half the Persepolis size, surrounded by trees. Note the Persepolis building (surrounded by a green rectangle) is elongated about 20° from the north. You might remember, I discussed earlier, that the Ziggurat of Chogha Zanbil seems to be tilted about 20° clockwise. The degrees are the same, which may surprise you.

Figure 11–Persepolis
Green Rectangle is added.
Google Earth Figure, North wise. Date 4/9/2018, eye alt: 8,730 ft.
IMAGE: CNFS/ Airbus

Figure 12 is a Google Earth map. It shows **Sha'Z Square** and the Persepolis ruins (Spot E). Persepolis is positioned almost in a continuous line that stretches from the Ziggurat of Dur (Spot A) and Chogha Zanbil (Spot C). The center of the yellow circle is at the Ziggurat of Chogha Zanbil (Spot C). I drew this circle to make the distance comparison easier.

Note how the Chogha Zanbil (Spot C) location has the same distances as the Ziggurat of Dur (Spot A) and Spot P.

This map would have been geometrically more excellent if Persepolis were on Spot P. This Spot is on the yellow circle's circumference.

And it is on the line passing from the diameter of A to C.

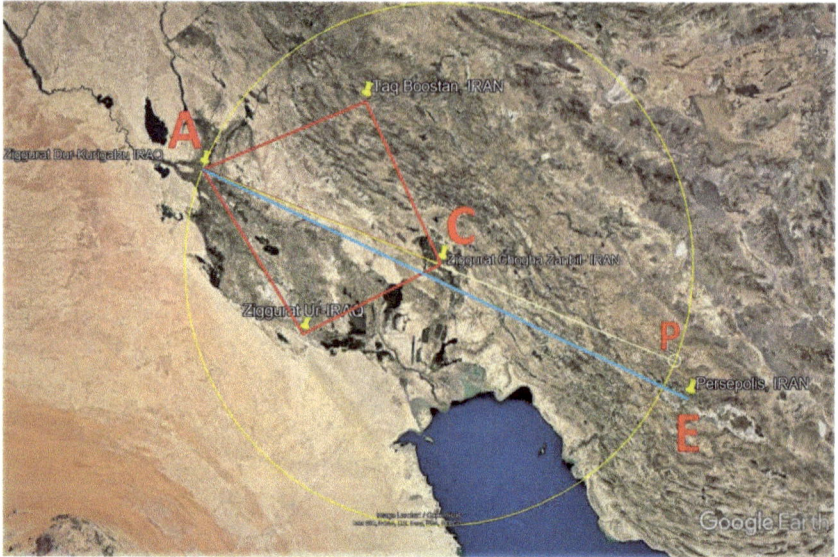

Figure 12–Sha'Z Square and Persepolis Colored lines, circles, and
letters are added
Google Earth Figure, eye alt 714.66 mi. North wise.
IMAGE: Landsat /Copernicus, DATA: SIO, NOAA, US NAVY, NGA, GEBCO

As shown in previous figures, it seems like our ancestors chose the
Spot for the structures carefully to correlate with each other geometrically.
Now, I want to see if there are other correlations in between.

The illustration below highlights the side-by-side comparison of
Sha'Z Square's shape with the other structures. Each of the smaller
figures is by Google Earth. Top-left would be the Ziggurat of Dur-
Kurigalzu. Topmiddle, Ziggurat of Ur Top-right you would see Ziggurat
of Chogha Zanbil and in the bottom left, the ruins of Persepolis. Lastly,
at the bottom-right is **Sha'Z Square** from above.

By comparing, you can tell how the shape and direction of the
structures are like **Sha'Z Square**. It is unlikely that the designers did not
know how to build their buildings elongated towards the northsouth
and west-east direction.

16

Figure 13–Illustration of comparison the shape of Sha'Z Square
(red squaredown right) with the other sites. From previous images.

Temple of Apollo

The next figure shows the **Temple of Apollo,** and some other structures located at **Patroos** in **Greece**. This artistic building is a cuboid shape with many right angles and many pillars in cylinder shapes and circular bases.

The ruined structure on the lower right side of the figure is impressive.

The red line is an indicator that is parallel with the long walls below it. You can easily see how the temple is not elongated parallel to the red line. Note that the red line is precisely 88° from the North and not a popular right angle of 90°. There is a 2° difference. Although the Greek designers knew the geographic east-west direction very well, they still made a small difference. It is unclear to me if they did it intentionally.

Figure 14–Temple of Apollo, Greece
Google Earth Figure, eye alt 1072 ft. North wise.
IMAGE: Maxar Technologies

The green line connects the Temple of Apollo to Persepolis *through Sha'Z Square*. You might want to know why the Temple of Apollo's direction is towards *Sha'Z Square*, besides, why the designers did not build it in the same order as its nearby structures.

The Temple of Apollo's design is different from the two Ziggurats mentioned earlier, but it is similar to Persepolis, which makes you think of an intercorrelation between them.

A Google Earth map of the wider area is in follow.

Spot G

Figure 15 is a Google Earth map. It displays the Temple of Apollo (Spot F), the Persepolis (Spot E), Dur (Spot A), Chogha Zanbil (Spot C), *Sha'Z Square*, and Spot G. The center of the red circle is Spot C, and its radius is to the Ziggurat of Dur (Spot A). It means the distance from A to C is the same as C to G. Note that the Spot G on this map is slightly lower than Spot P, in Figure 12.

The yellow line starts from Spot F, almost crosses over Sha'Z Square's diameter, connects to Spot C, and reaches the circle in Spot G.

It seems like the designers made a miscalculation to build the Persepolis in Spot E. You can see how geometrically interesting it would have been to build the Persepolis on Spot G versus Spot E.

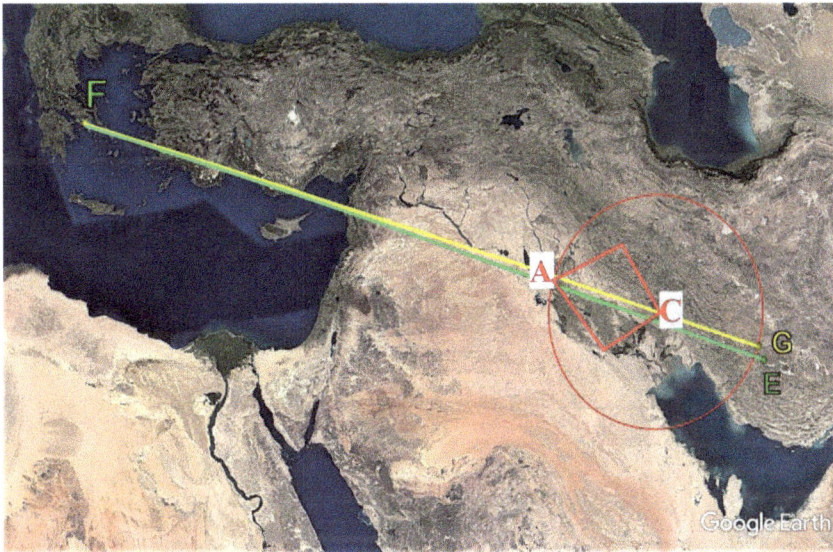

Figure 15–Temple of Apollo, Sha'Z Square, and Persepolis
Google Earth Figure, eye alt 1657.77 mi. North wise.
IMAGE: Landsat /Copernicus

Let us study this map in more detail. The next figure represents a closer look at **Sha'Z Square**, Persepolis, and Spot P. The green line is from the Temple of Apollo that reaches Persepolis in Spot E. The yellow line is also coming from the Temple of Apollo. It passes from above Spot A, reaches Spot C (the Ziggurat of Chogha Zanbil), then continues to intersect with the red circle within Spot G. Spot G is slightly lower than Spot P.

Spot G's latitude is about 30.408173°, and if rounded to the nearest tenth place becomes 30.4°. The number 4 in the tenth place puts Spot G into Group 2- refer to ***ancient coordination system*** in part 1.

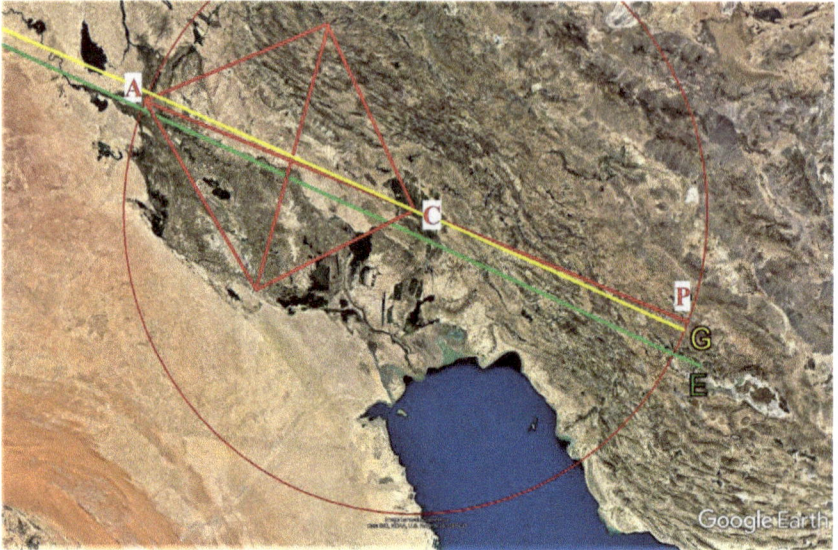

Figure 16–Sha'Z Square, Persepolis, Spot P, and Spot G
Colored lines, red circle, letters, and names are added.
Google Earth Figure, eye alt 653.53 mi. North wise.
IMAGE: Landsat /Copernicus
DATA: SIO, NOAA, US NAVY, NGA, GEBCO

I did this map to understand the position of the Spots better. You might draw some lines and circles and come up with more exciting results. Interestingly, there are more structures in the area, some of which are between Spot E and Spot G. I will study one of them later in this article.

Regardless of the small tweaks, our ancestors chose specific spots for their artistic structures.

Another note to consider is that the shape and elongation of the buildings, including the direction of the Ziggurats' steps, suggests they might have acted as direction signs. They look like our lighthouses in the oceans or the street signposts for warnings and directions.

I will present some studies in the next sections about other locations to see more exciting discoveries.

Part 2
Kaaba of Mecca

There is a big cubic-shaped structure by the name Kaaba located at **Mecca** city in **Saudi Arabia. Kaaba** is an Arabic word that means "cube" in English. According to many ancient articles, the Kaaba structure was a holy site for various tribes of the territory. Today, it is the most sacred site in **Islam**. According to **Andrew Peterson**, in the Dictionary of Islamic Architecture, Muslims performing their Islamic prayer from any part of the world are expected to face its location.

The origin of Kaaba is vague for many people. The **Quran** (the central religious text of Islam) contains several verses that say Kaaba is the first house appointed for humanity and Ibrahim and Ishmael once raised its foundations. Two of the verses are as follow:

- (Quran, Chapter 3 "**Aale-Imran**" verse 96):

    "The first House (of worship) appointed for men was that at Bakka: Full of blessing and guidance for all kinds of beings." [4] [The Meanings of the Holy Qur'an by Abdullah Yusufali]

    I wonder if the word "guidance" means geographic direction.

---

4.   http://www.islamicity.com/mosque/quran/3.htm

- (Quran, Chapter 2 "***Al-Bakarah***" verse 127):

    And remember Abraham and Isma'il raised the foundations of the House (With this prayer): "Our Lord! Accept (this service) from us: For Thou art the All-Hearing, the All-knowing." [5] [The Meanings of the Holy Qur'an by Abdullah Yusufali]

Besides the Quran, there are ***Hadiths***, which are something unproven but attributed to the Islam prophet that is not in the Quran. Some Hadiths say Kaaba is dated back to ***Adam's*** time (the first man) or even before him. ***Ibn Kathir*** (famous commentator on the Quran) mentioned that the shrine (Kaaba) was a place of worship for angels before the creation of man and later, was lost during the flood in ***Noah's*** time and then it was built on the location and was finally rebuilt by ***Abraham*** and ***Ishmael***. [6]

You can read verses such as 14:35-41, 2:125-129, and 22:26-27 from the Quran to study more.

Qiblah

In geometry, a vertex is where two or more curves, lines, or edges meet. Kaaba structure is not exactly a geometrical cube with the same vertexes.

According to ***William Montgomery*** Watt in his book, Muhammad prophet and Statesman, For Muslims, the direction for prayer and burial from any spots on Earth is called the Qiblah. The Qiblah of Islam was the Al-Aqsa Mosque located in the Old City of Jerusalem from 610 CE for about thirteen years. In 624 CE, ***Mohammad*** (the prophet of Islam) changed the direction of Qiblah towards Kaaba.

Geographical Coordinates of Kaaba

The Geographic Coordinates of Kaaba is in the next table.

---

5.   https://sacredtexts.com/isl/quran/00215.htm
6.   https://slife.org/kaaba/

Group	Name	Location	Coordinate (Decimal)	Latitude (rounded)
2	Kaaba	Saudi Arabia, in Mecca	21.422500° 39.826184°	21.4°

Table 3–Geographic Coordinates of Kaaba and its rounded latitude

Note how the last digit of the rounded latitude in the nearest tenth place ends with the number 4, which puts it in Group 2.

By comparing this latitude number with the ones in Table 2 of Part 1, you might wonder if there are unspecified correlations among them.

It is more apparent now that the direction and the elongation of ancient buildings and structures were most important. Most utmost mosques are faced or elongated towards Kaaba, and they have a wall niche inside the building that indicates the Qiblah. But what about the structures and streets before Islam? This article studies the direction of the sites regardless of their religion and astronomical interests.

Origin

Although some historians have doubts about the origin of the Kaaba, if it is true, then Kaaba is the oldest structure on Earth and an important archaeological and historical spot for humankind.

Kaaba and Sha'Z Square

The next figure is a comparison of two places from above and north wise. The top picture is Kaaba. The white roof of Kaaba has a different orientation with the geographical north-south direction. The lower image is *Sha'Z Square*, and the red lines are its borders. The same shape of the red lines of *Sha'Z Square* from the lower picture is added to the top image near the roof of Kaaba to compare shape and direction. As shown in Figure 17, the order and the form of both *Sha'Z Square* and Kaaba's top in scale are nearly alike.

You can refer to Figure 13 to compare Kaaba's roof's shape and direction with other structures mentioned earlier. Yes! Another shocking discovery.

This similarity is another correlation, and of course, it might raise more questions again. You might even want to go back and start reading this article from the beginning to review everything a bit deeper. You might ask yourself questions such as:

Did our ancient designers know about them all? If they did, then how?

Figure 17–Comparison of Kaaba's shape with Sha'Z Squareadded red shape
Top: Kaaba's roof. Figure IMAGE: Maxar Technologies
Google Earth Date: 2/7/2019, eye alt: 1249 ft. North wise.
Bottom: Sha'Z Square. IMAGE: Landsat /Copernicus
Google Earth Date: 12/30/2016, eye alt: 1410.24 mi. North wise.

Trigonometry

So far, in this article, we saw how our ancient designers used mathematics. They have used basic geometry such as lines, rectangles, squares and circles for their locations in their two- dimensional structures and maps.

We also noticed how they used basic three-dimensional shapes such as simple rectangular prisms, quadrilateral and cuboids in their structures. But how about the usage of more complicated shapes such as triangles and pyramids?

There are three popular triangles:

- The *equilateral triangle* has three congruent sides and three equal angles of 60° each.
- The *isosceles triangle* has two equal sides and two equal angles.
- The *scalene triangle* has different lengths and different angles. The most popular scalene triangle is a right triangle, which has a 90° angle in it.

The 30-60-90-degree triangle is the only right triangle whose numbers are in an arithmetic progression. And they are common numbers in geometry, especially in trigonometry.

Some popular angles in geometry are 30°, 45°, 60° and 90°. Within some parts of this article, you will see ancient examples of these angles and shapes.

In the next part, I represent the Taq-e Bostan site to find more footprints regarding a mathematical interest in the location.

Part 3
Taq–e Bostan Triangle

Although Taq-e Bostan was from the 4th century A.D., before the rise of Islam, I wanted to show other geometric correlations with the location of Kaaba regardless of any religion or astronomical interests.

Near the Taq-e Bostan site, located in Kermanshah, Iran, there is a straight road called *Bolvar– e – Taq-e Bostan*. This old road was essential because it was the primary way that locals used to visit the site. Today it is called *Bolvar – e – Shahid Shirudi*. It got renewed about 50 years ago, and it is said to be the longest boulevard in Iran.

The next figure shows a map of the Taq-e Bostan area by Google Earth.

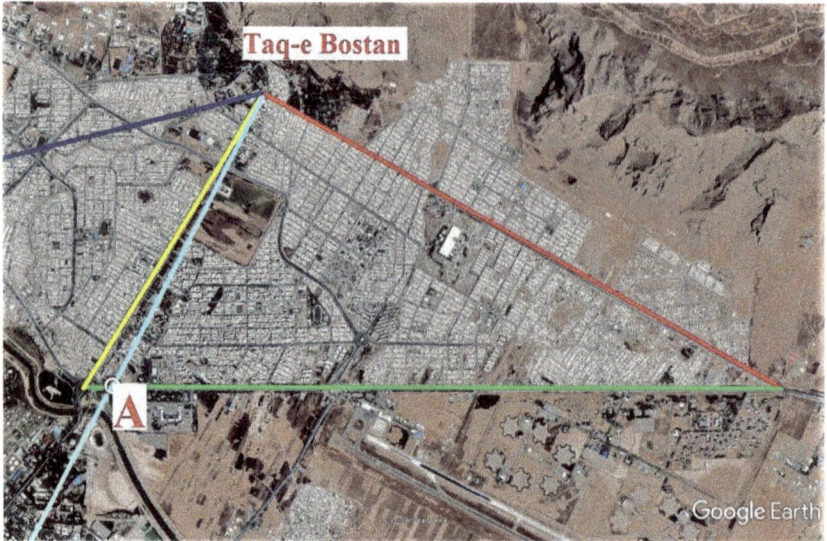

Figure 18–Taqe Bostan Triangle
Colored lines, letters, and letters are added.
Google Earth Figure date 7/19/2019, eye alt: 29433 ft. north wise.
IMAGE: CNFS / Airbus

The yellow straight line is precisely on the Bolvar – e – Shahid Shirudi road. It is about 32.20° from the north and about 2.24 miles long. The green straight line is another old road about 4.64 miles, entirely in the eastwest direction. The red straight line is about 3.93 miles. It is on another old road that intersects with the green line and creates a vertex with a 28.56° angle.

This angle is only 1.44° less than the popular 30° in triangles.

The yellow, green and red lines create a triangle shape, which is why I chose this site to study. This triangle area is about 4.35 mi².

There are some grayish houses and an airport in the territory. You can see how aligned they are with the mentioned lines.

Now, let us check the angles and the vertexes of this Triangle. The angle between the yellow and red line is 93.60°, which is just 3.60° more than the popular 90° in a right triangle.

The light blue straight line is from Kaaba to the intersection of the yellow and the red lines; the Northern corner of this big Triangle.

There is a 3.76° difference between the yellow and the light blue line. We can omit this due to the 4.35 mi² area. It might be another minor miscalculation.

Interestingly, the light blue line has 89.84° with the red line. This angle is only 0.16° less than a popular 90° in a right triangle. In other words, the light blue, red and green lines almost create a right triangle with each other. It means if the designers made only one small correction and built the boulevard on the light blue line instead of on the yellow line, they would have had created almost a perfect 30-60-90-degrees right triangle-shaped area with a boulevard precisely towards Kaaba.

First, they needed to change the road direction for just 3.76° from the Triangle's top corner to shift the yellow line road to the light blue line shown in the Figure. The boulevard turns towards Kaaba. Secondly, they needed to begin the green line from the intersection with the light blue line at spot A, making the green line about 0.18 miles shorter.

The dark blue line attached to the big Triangle's top corner is coming from the Giza Pyramids. The angle between this dark blue line and the light blue line on the Triangle's top is about 48.09°, only 3.09° more than the popular 45° in triangles. This number is another interesting angle. Almost a 3° difference in about 1,000 miles from Taq-e Bostan to Giza is another minor miscalculation in that era that the designers did that can be omitted.

Anyways, with only about 3° or 4° differences, the Persian designers in the fourth century could have created a perfect 30-60-90-degrees right triangle area with a road directed towards Kaaba.

The importance of Kaaba and Giza Pyramids to Persians in that era is unclear to me. Persians already had some interactions with Egyptians before the 4th century that might have created some interests regarding the Giza Pyramids. Still, Islam originated later at the beginning of the seventh century C.E.. What is clear to me, the two Ziggurats of Chogha Zanbil and Tepe Sialk are similar to a stepped pyramid shape structure in Egypt. besides, the Persians never built an important complete pyramid shape structure before. Instead, they made a vital cuboid structure called Kaaba Zartosht before the Islamic era. I do not know the interest in building a different shape structure yet.

The next section is about the Kaaba Zartosht structure.

Part 4
Kaaba Zartosht

Our ancient engineers built the **Kaaba Zartosht** or '**Ka'ba-ye Zartosht'** in the First **Persian Empire** era (the **Achaemenid Empire**) about 550– 330 BCE near the ruins of Persepolis in Iran. It was called **Bon-Khanak** in the **Sassanian** era; the structure's local name was **Kornaykhaneh** or **Naggarekhaneh**. People use the phrase Ka'ba-ye Zartosht from the 14t^h century into our contemporary age. [7.]

It is between the Spots E and G in Figures 15 and 16. One exciting thing about this site is its cuboid shape. The next Figure shows the Kaaba Zartosht from the side.

The main reason that Persian designers used a cuboid shape versus other shapes to build this Kaaba Zartosht is unknown to me, but they could have made it in any form and any location.

The geographic location of this place is in the next table:

---

7.   https://en.wikipedia.org/wiki/Ka%27baye_Zartosht 1/9/2020

Group	Name	Location	Coordinate (Decimal)	Latitude (rounded to tenth)
1	Kaaba Zartosht	Marvdasht, Iran	29.988889° 52.874722°	30.0°

Table 4 Coordinates of Kaaba Zartosht

Note the nearest rounded latitude digit to tenth place is *0*, which places this site in Group 1.

The location is closer to the Spot G in Figures 15 and 16 rather than the Persepolis ruins, which I discussed earlier. Besides, the structure is unique in the region, so you might think perhaps Persian engineers wanted to correct their miscalculations.

You might also wonder why they used a cubic shape similar to Kaaba and not another form, such as a Square-Based Pyramid similar to Egyptian Pyramids. Also, note that the building has triple steps in front of its only entrance. There are multiple squares nested in each other around it. You can see a similar design as the Ziggurats above and Kaaba.

Nevertheless, the design of Kaaba Zartosht has always kept its glory and kudos ever since its creation.

It is unclear what the designers wanted to demonstrate. I will refer back to this site later again.

Figure 19–Kaaba Zartosht (Ka'baye Zartosht), Iran
Attribution: By Diego Delso, CC BYSA 4.0
https://commons.wikimedia.org/w/index.php?curid=52061185

Part 5
Taq-e Bostan's Equidistance

Now that it is clearer how some of our ancestors were interested in Kaaba's location and shape, I am going to open a broader map and study this matter from a wider perspective. I want to know if there were any interest in the Giza Pyramids' location and Kaaba's.

The next table shows how the distances from Taq-e Bostan to Kaaba and Giza Pyramids are almost equal (only about 13 miles difference).

From Taq-e Bostan	Distance (miles)
Kaaba	997.59
Giza Pyramids	984.40
Difference	**13.19**

Table 5–Distances from Taqe Bostan to Kaaba and Giza Pyramids

The next Figure shows a Google Earth map of Egypt, Saudi Arabia, and Iran from above and North wise. Kaaba is in the Southern part of the Figure.

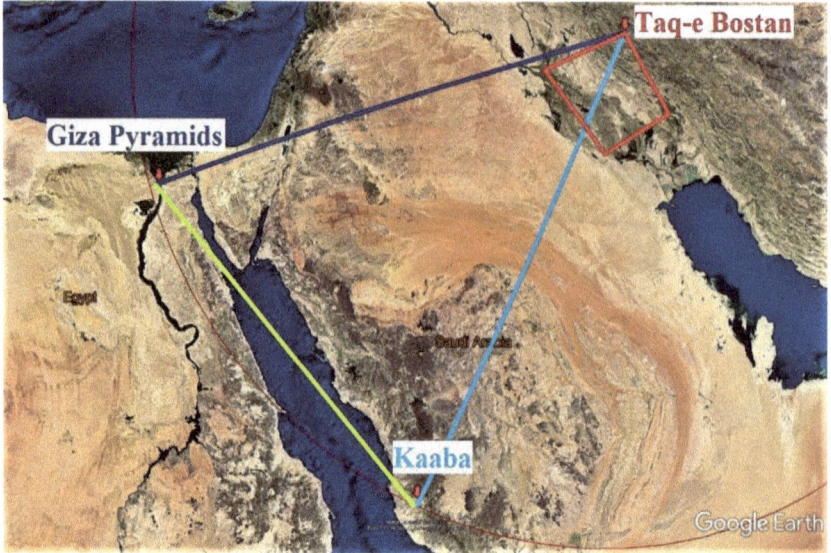

Figure 20–Taqe Bostan's equidistance with Kaaba and Giza Pyramids
Google Earth Figure date 12/2016 eye alt: 1389.91 ft. north wise.
IMAGE: Landsat /Copernicus

The red Square is **Sha'Z Square**, where Taq-e Bostan is in the Northern corner of it discussed earlier in this article. The light blue line connects the Taq-e Bostan to Kaaba, and the dark blue line connects the Taq-e Bostan to the Giza Pyramids, which was discussed earlier and presented in Figure 18. The green line is from Kaaba to Giza Pyramids. The Figure shows a portion of a red circle. The center of it is Taq-e Bostan, and its radius is to Kaaba with about 997.59 miles. All the areas underneath the circumference of the red circle have the same distance to Taq-e Bostan.

Note the small gap of about 13.19 miles from the Giza Pyramids to the red circle. Refer to Table 5. This small gap is unclear, and it might have been due to miscalculations of the Taq-e Bostan's location. Or perhaps it was due to a ground-shift of the area.

At first glance, the straight lines create a shape that looks like a big Isosceles Triangle with two equal sides and two equal angles. You might ask yourself if it is a coincidence that the Taq-e Bostan has almost the same distance as Kaaba and Giza Pyramids. But, how about their angles with each other?

First, let us look at the big Triangle from the Taq-e Bostan perspective. The angle at the Triangle's top, Tag Bostan corner, between the light blue line and the dark blue line, is about 48.09°. This number is only 3.09° more than the popular 45° angle. It is almost half of a right angle.

Second, let us look from the perspective of Kaaba and Giza Pyramids sites towards the Taq-e Bostan. The dark blue line from Giza Pyramids to Taq-e Bostan is 67.79° from the North, 22.21° from the East. The blue line from Kaaba to Taq-e Bostan is 24.83° from the North, 65.17° from the East. These are almost mirror numbers. The angles and their differences are in the next table.

To Taq-e Bostan	North (degree	East (degree)
Kaaba	24.83 °	65.17°
Giza Pyramids	67.79 °	22.21°
Difference	42.96 °	43.01°

Table 6–Directional angles from Taqe Bostan to Kaaba and Giza Pyramids

The differences are 42.96° and 43.01°, which are remarkably close to each other. This tiny difference of 0.05° in their calculations is nothing compared to the thousand miles distance.

The big triangle angles are as follows: Taq-e Bostan corner is 48.08°, Kaaba is 65.51°, and Giza Pyramids is 67.74°. The vertexes of Kaaba and Giza Pyramids corners are almost similar (only 2.23° difference).

These numbers mean the big Triangle is remarkably identical to an Isosceles Triangle with two equal angles. It is hard to believe they are a coincidence. You might wonder what the purpose of creating this big triangle shape is, which can only be seen from the sky regardless of the direction on Earth's surface. How and why, this was designed and used is still a mystery.

The reason for the small differences is unknown yet. According to the design of *Sha'Z Square* and the earlier building of Kaaba and the Giza Pyramids, I think of two possibilities. The engineers made a small miscalculation or did not have a better locational choice in the mountains. Finding the truth requires more study.

Further down this article, you will see some other exciting footprints, such as the different ancient Egyptian designs.

Part 6
Colossi of Memnon

Scientists and scholars have learned from Egyptians that they did not randomly go through many efforts for their constructions. In this part and later again in this article, I will show some footprints of how some ancient designers were interested in building their structures and monuments in specific directions.

There are two massive stone statues of the *Pharaoh Amenhotep III* in Egypt, who reigned during *Dynasty XVIII*. Since 1,350 BCE, they have been placed in the *Theban Necropolis*, located West of the *River Nile* from *Luxor's* modern city at the coordinate of 25.720469°, 32.610476°.

Note that the nearest tenth place of its rounded latitude is seven, which is not in the range of Groups 1 or 2, discussed earlier in Part 1 of this article. Perhaps it can be in a new one like Group 3.

The original function of the Colossi was to stand guard at the entrance to *Amenhotep's* memorial temple (or mortuary temple): a massive construct built during the Pharaoh's lifetime, where his people worshipped him as a *God-On-Earth* both before and after his departure from this world.[8]

---

8.   Wilfong, T.; S. Sidebotham; J. Keenan; DARMC; R. Talbert; S. Gillies; T. Elliott; J. Becker. "Places: 786066 (Memnon Colossi)". Pleiades. Retrieved March 22, 2013.

The twin statues depict **Amenhotep III** (fl. 14th century BCE) in a seated position, his hands resting on his knees and his gaze facing eastwards (actually **ESE** in modern bearings) towards the river. There are two shorter figures carved into the front throne alongside his legs: his wife **Tiye** and mother **Mutemwiya**. The side panels depict the **Nile god Hapi**.[9]

Figure 21–Colossi of Memnon

The photo was taken by Hajor, December 2001 and released under the Creative Commons Attribution-Sharealike License and the GNU Free Documentation License.

But if you see carefully, the guardian statues are sitting on a cubic chair, similar to the Taq-e Bostan King in Figure 10.

You might ask, why were these gigantic guards not built standing straight on their feet? Why on cubic chairs? Is there any correlation between these two guards and **Sha'Z Square** or Kaaba?

9. https://en.wikipedia.org/wiki/Colossi_of_Memnon - 1/9/2021

The next Figure is by Google Earth, showing the structures from about the North-East side. I added the two 3D models of the monuments, the name, and the colored lines to the Figure by Google Earth's featured tools.

The yellow line indicates the geographic north direction from the middle of the two structures. In other words, it has $0°$ from the North. I drew the red line from the middle of two monuments to the center of Kaaba by Google Earth program. It is about $121.53°$ from the North. In that era, the Egyptians were completely aware of the geographic North-South and could build these two gigantic structures towards any desired direction.

Figure 22–Colossi of Memnon front
Google Earth Photorealistic tools add 3D monuments.
Google Earth Figure date 9/18/2019, eye alt: 567 ft., about North East angle.
IMAGE: Maxar Technologies

You can also see how the red line is almost parallel with the road on the Figure's left side. Like the Bolvar – e – Shahid Shirudi road in Taqe Bostan area (refer to Part 3), this road is almost elongated towards Kaaba.

According to this Figure, it is obvious that the two guardian statues are facing Kaaba. This study brings up many other questions, such as:

- Does the number 121.53 mean something special to them?
- Is this direction an ancient signpost?
- Were these statues accidentally positioned towards Kaaba?
- Why have they chosen to sit on cubic chairs?
- Is there other Egyptian evidence of Kaaba-directional structures?

Part 7
Giza Pyramids and the Sphinx

Here I want to study a more prominent site in the territory, perhaps the biggest one. According to archaeologists, the Pyramids of Giza include the *Great Pyramid of Khufu* (right), the *Pyramid of Khafre* (middle), and the *Pyramid of Menkaure* (left), along with their associated pyramid complexes and the *Great Sphinx* (lower right).

Figure 23–Giza Pyramids and the Sphinx
Google Earth Photorealistic tools add 3D monuments.
Date: 11/2018 eye alt 2514 ft. Almost North East angle
IMAGE: Maxar Technologies

Egyptians constructed the pyramids between 2,580 BCE and 2,560 BCE during the fourth Dynasty of the *Old Kingdom* of *Ancient Egypt*. Antipater of Sidon listed the Great Pyramid as the oldest *Seven Wonders of the World*, the only one still in existence, and the tallest human-made ancient structure in the world.

The Sphinx in the South-East of the Khufu Pyramid is one of the world's largest and oldest statues. Although its construction's exact date is unknown, some believe it was in approximately 2,500 BCE, according to "**Sphinx Project: Why Sequence is Important**" (2007).

The next Figure shows an older picture of the Sphinx and the Khufu Pyramid in Egypt before the clearance.

Figure 24–Old picture of the Khufu pyramid in the back and the Sphinx in front
https://commons.wikimedia.org/wiki/File:The_Great_Pyramid_and_the_Sphinx.jpg
Attribution: Scottish National Gallery [Public domain]

Giza Pyramids Geographical Coordinates

Some information about the three main Giza Pyramids and the Sphinx are on the next table:

Group	Name	Location	Approximate Build Time	Latitude (Degrees)	Latitude (rounded)
1	Khufu	Giza in Egypt	2,580-2,560 BCE	29.979147°	30.0°
	Khufu			29.975997°	30.0°
	Menhaure			29.97.2497°	30.0°
	Sphinx		Unknown	29.975263°	30.0°

Table 7–Some Giza Pyramids and Sphinx information

Note that the nearest tenth place of each of their rounded latitudes is 0. This number adds the sites to Group 1. As a reminder, the round number to their nearest tenth place of Persepolis and Kaaba Zartosht latitudes are also 30.0°, which you might wonder if there is a correlation.

Giza Pyramids geometric design

The next Figure is from the Google Earth map from above towards the North. It quickly shows the orientation of the Giza Pyramids geologically to the North-South and East-West direction. The yellow lines and the white letters are guides added to the Figure to understand the article better. The big yellow squares are identical in size. The Red Circle center is at the center of the Khafre Pyramid and radiuses to the North-East corner of the Khufu Pyramid (Spot A). Note how the Red

Circle reaches the Sphinx structure located at the Eastern side of the Red Circle.

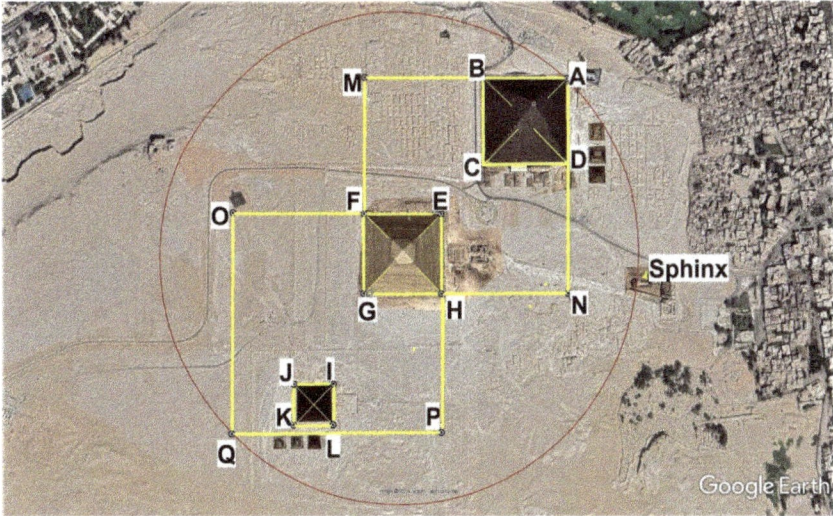

Figure 25–Giza Pyramids and Sphinx area
The 3D buildings, colored lines, red circle, and letters are added.
Google Earth Figure from above, eye alt 6624 ft, North wise
IMAGE: Maxar Technologies

Google Earth measures the yellow lines' length and direction of each Spot's coordinates in the next tables.

The next Figure shows a more detailed correlation of the previous Figure. The Red Circle center is at the Khafre Pyramid center at Spot CtrB (middle Pyramid). The green line of ACEGQ is the Red Circle's diameter, which almost crosses the Khafre and Khufu pyramids' diameter. The two purple lines of BFR and DHZYS connect the NorthWest and the South-East corners of the Khafre and Khufu pyramids and are almost parallel with the green diameter line. Note how the purple line from D to H crosses the ruins on the right side of the Khafre Pyramid. This structure is called "*the Funerary Temple of Khafre*," and its front section has a square shape, which this purple line passes over its diameter. You can find more correlations there.

49

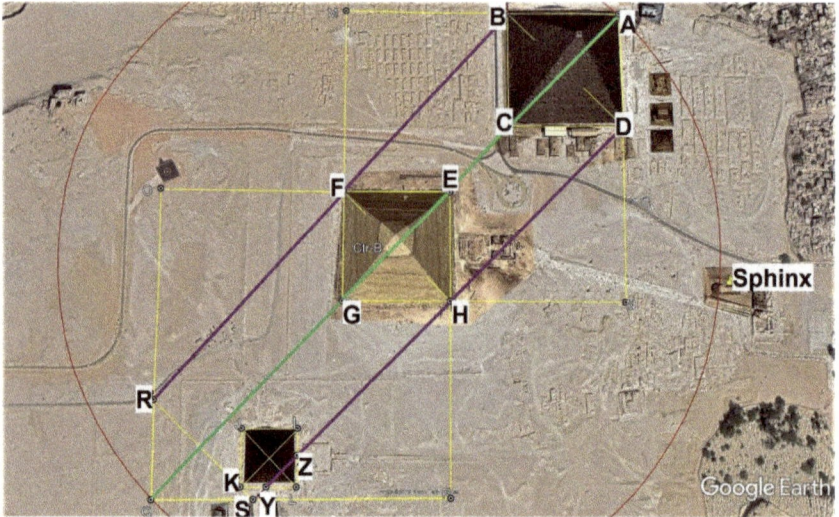

Figure 26 Giza Pyramids and Sphinx areacloser
The 3D buildings, colored lines, red circle, and letters are added
Google Earth Figure from above, eye alt 4645 ft. North wise
IMAGE: Maxar Technologies

Giza Pyramids diameters towards Kaaba

The next Figure is from the Google Earth Map of the Giza Pyramids from above, including some more geometrical correlations.

There are four green lines added to the map that is from Spots B, E, F, and J to the center of Kaaba (Kaaba is not on this map due to longdistance). This geometry is exciting. You might think of a correlation in between.

Three of the green lines pass the three pyramids' diameter, and the green line from Spot E goes through the diameter of the ***Funerary Temple*** of ***Khafre*** (red Square). It seems like they had precisely calculated the Funerary Temple's shape and Spot for this purpose. These precise geological alignments to North-South and their pyramid diameters towards Kaaba in such immense structures do not seem random.

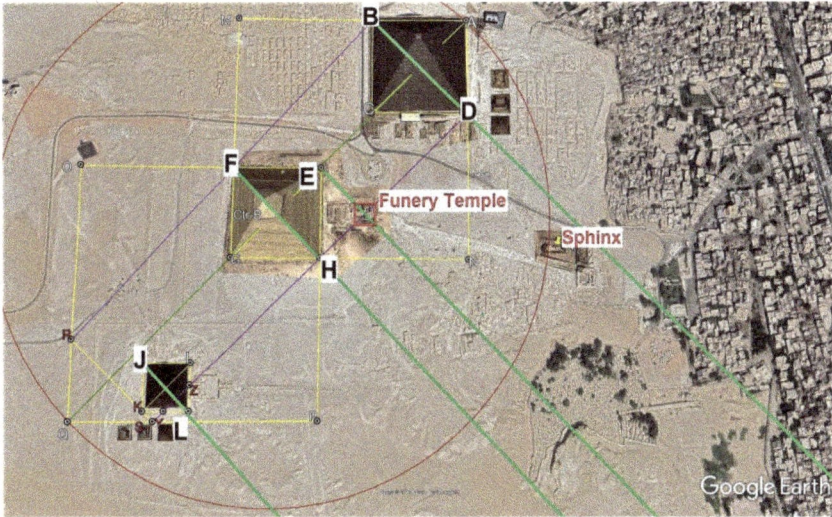

Figure 27–Giza Pyramids and Sphinx area direction to Kaaba
The 3D buildings, colored lines, red circle, and letters are added.
Google Earth Figure from above, eye alt 5559 ft. North wise
IMAGE: Maxar Technologies

By analyzing deeper, you can find more geometrical correlations. The diameters of the three main Pyramids and the two Colossi of Memnon monuments from the previous section do not seem to be built towards Kaaba accidentally.

You might find more interesting geometric correlations in this matter. Later, I will show another group of ancient pyramids that their diameters also play the same role, with this difference that they are older and further.

Table of the connection-lines

The next two tables represent some useful information. They show the approximate ground length of the lines used on the last figures, and they show the direction. The measurements are by the Google Earth features.

Lines	Direction (Degree)	Ground Length (Meter)
D-AC	224.82°	323.9
D-BD	135.25°	323.9
D-EG	225°	294
D-FH	134.90°	294
D-IK	225.15°	148
D-JL	134.96°	148
L-AB	270°	229
L-AG	223.89°	799
L-BC	180°	229
L-BFR	222°	1009
L-BM	270°	326
L-CD	90°	229
L-DA	0°	229
L-DHY	45.61°	961.2
L-EF	270°	208
L-FG	180°	208
L-FO	270°	352
L-GH	90°	208
L-HG	0°	208
L-HN	90°	347.6
L-IJ	270°	105
L-JK	180°	103
L-KL	90°	105
L-LI	0°	104
L-MF	180°	361
L-ND	0°	338.3
L-OQ	180°	563.5
L-PH	0°	358
L-QP	90°	560

L-RKS	133.26°	259.2
Ln-H1-B	135.56°	1,287,452.5
Ln-H1-F	135.53°	1,287,426.2
Ln-H1-J	135.51°	1,287,244

Table 8–Ground length and the direction of the lines

Spot	Latitude		Longitude	
	Degree	Decimal	Degree	Decimal
A	29°58'48.67"N	29.980186°	31° 8'7.46"E	31.135406°
B	29°58'48.67"N	29.980186°	31° 7'58.91"E	31.133031°
C	29°58'41.21"N	29.978114°	31° 7'58.91"E	31.133031°
D	29°58'41.21"N	29.978114°	31° 8'7.46"E	31.135406°
E	29°58'36.96"N	29.976933°	31° 7'54.53"E	31.131814°
F	29°58'36.96"N	29.976933°	31° 7'46.75"E	31.129653°
G	29°58'30.23"N	29.975064°	31° 7'46.75"E	31.129653°
H	29°58'30.23"N	29.975064°	31° 7'54.53"E	31.131814°
I	29°58'22.70"N	29.972972°	31° 7'43.75"E	31.128819°
J	29°58'22.70"N	29.972972°	31° 7'39.83"E	31.127731°
K	29°58'19.34"N	29.972039°	31° 7'39.83"E	31.127731°
M	29°58'48.67"N	29.980186°	31° 7'46.75"E	31.129653°

N	29°58'30.23"N	29.975064°	31° 8'7.46"E	31.135406°
R	29°58'24.36"N	29.973433°	31° 7'33.68"E	31.126022°
Y	29°58'19.34"N	29.972039°	31.7'41.63"E	31.126022°
L	29°58'19.34"N	29.972039°	31° 7'43.75"E	31.128819°
Z	29°58'21.15"N	29.972542°	31° 7'43.75"E	31.128819°
O	29°58'36.96"N	29.976933°	31° 7'33.68"E	31.126022°
P	29°58'18.63"N	29.971842°	31° 7'54.53"E	31.131814°
CtrA	29°58'44.93"N	29.979147°	31° 8'3.19"E	31.134219°
CtrB	29°58'44.93"N	29.975997°	31° 7'50.64"E	31.130733°
CtrC	29°58'20.99"N	29.972497°	31° 7'41.80"E	31.128278°

Table 9–The coordinates of the spots on the map

Sphinx vs. Kaaba

The next Figure shows a Google Earth image with a closer look at the Sphinx area from above. The **Sphinx** is at the lower right section. I added the letters and names, the lines, and a portion of the Red Circle to understand the contents.

Figure 28–Sphinx and Southern part of the Khufu pyramid.
I added the yellow and green lines, the red circle, and the red and white letters
Google Earth Figure 10/26/2018 eye alt: 1976 ft. from above, North wise.
IMAGE: Maxar Technologies

There are two green lines, which are from Spot U and Spot W to Kaaba. Spot U is on the Southern face of the Khufu Pyramid in about one-fifth of the total height. You can see this Spot on the Pyramid wall in Figure 24, which seems like a damaged section.

Spot W is in the middle of the Southern edge of the Khufu pyramid.

Note how the green line from spot W intersects the Red Circle on the back of the Sphinx, and the green line from Spot U crosses the Sphynx head.

With everything said, I do not think the Sphinx is built in that location by accident, and they create it in that size and direction on purpose.

Location vs. shape

So far, we have noticed how the diameters of the pyramids and the Sphinx location show some interest among the Egyptian designers towards the Kaaba location. Still, it is not the other way around. I mean, none of the diameters of the Kaaba reaches to Giza Pyramids. The reason is unknown to me, but it might be due to age.

Perhaps because Kaaba was an older structure and known to the Egyptians to build their Pyramids in a shape, location, and orientation to create some respect, or it might have been entirely due to something else. More study is required.

You might ask why the Egyptian designers took a lot of effort and built their gigantic structures in a Pyramid Shape instead of a simple cubic shape. You might also want to know why they decided to use a more complex geometric form with different angles and vertexes for their magnificent structures while building their ordinary houses in rectangular prisms or cuboid shapes with right angles.

In the next section, I am peeking into the shape of the structures with more detail.

Pyramids vs. Cubes

In Egypt, imagine what would have happened if there were cylinder, cubic or triangle-based pyramid structures instead of the squared-base pyramid ones. What would have happened if Kaaba was in another shape, such as a pyramid or a cylinder. One thing to note was their edges would have been different.

I understand that pyramids have fewer edges than cubes and that the designers might have been worried about edge erosions for their structures. I also know that the square-based pyramids have better stabilization on the ground than the cylinder or cube ones. But Egyptian engineers in that era could have built dome-shaped structures with rounded bases if their intention were stabilization. I mean something like igloo structures that Eskimo people built.

So, what was the main reason to use pyramid shapes? Perhaps, the Egyptians needed some stabilized directions as tall signposts in those days to see from a distance, and the big and tall pyramids gave them what they needed. The edges and the diameters of pyramid structures were some directions to use. They used square-based pyramid shape structures to provide them with the edges as the four North, South, West and East directions for navigation. That is what I think they needed. They have aligned the square base exactly North-South and East-West wise for geographic trends besides their intention towards the Kaaba spot. If the Egyptians built the tall Giza Pyramids somewhere in a higher or lower location by keeping its alignments precisely towards North-South, its diameters did not show the exact direction of Kaaba.

Since the **Pyramid of Djoser** time, Egyptians use the square-based Pyramid, even in their smaller sized pyramids. The Pyramid of Djoser is one of the earliest pyramid structures built-in 2,630–2,610 BCE. So, I think Kaaba's 'location' was crucial than its shape. Perhaps in that era, either the Egyptians were unaware of the Kaaba's structure, or the Kaaba structure did not exist. If that is true, then why was the location important?

Mathematics

This section wants to study how the ancient Egyptian architectengineers produced pyramid shapes for their magnificent structures. The pyramid structures are indications of the sophistication of Egyptian mathematics. But let us go back in time and try to understand their designer's thoughts better.

The **Ancient Egyptian Numeral System** was from around 3,000 BCE before building the Pyramid of Djoser. The system was numeration based on multiples of ten, often rounded off to the higher power. They wrote in hieroglyphs. The oldest mathematical text from ancient Egypt discovered so far is the Moscow Papyrus, which dates from the Egyptian Middle Kingdom around 2,000 – 1,800 BCE. Therefore, you might want to ask how the Egyptian engineers utilized mathematics and geometry to build their structures in that era. Numbers 4, 5, and 8 can be the keys to this question. These numbers have commonalities between the Pyramid shape and the number of classical elements in the ancient era. It sounds interesting.

The square-based Pyramid has a base with four corners and four edges, four equilateral triangles sides, five vertices and eight edges. So, there are three numbers of four, five and eight involved. Now, let us take a more indepth look and see if we can find some correlations in between.

A squared based pyramid is a **Pentahedron** because it has five faces: four triangle faces and one square face. Perhaps the key is in classifying the material world in ancient times with four or five elements.

According to *"The Physics of Plasmas" by Boyd, T.J.M.; Sanderson, J.J. (2003),* the classification of the material world in ancient time, including Hellenistic Egypt, was either four or five elements as *Air, Earth, Fire* and *Water,* and then *Aether* (was added later). For example, in Babylonian mythology, the cosmogony called *Enûma Eliš,* a text written between the 18th and 16th centuries B.C., involves four gods that we might see as personified cosmic elements: *Sea, Earth, Sky,* and *Wind.* [10], [11]

In India, there are five elements in Hinduism as *Earth, Water, Fire, Air* or *Wind,* and *Space* or *Aether,* and there are four elements in Buddhism as *Earth, Water, Fire* and *Air.* [12]

The Chinese had a somewhat different series of elements, namely *Fire, Earth, Metal* (literally gold), *Water* and *Wood,* which were as other types of energy in a state of constant interaction and flux with one another, rather than various kinds of material.

Feng Shui, also known as Chinese geomancy, claims to use these five energy forces to harmonize individuals with their surrounding environment.

But in Egypt, a text written in *Hellenistic* or Roman times called the *Kore Kosmou* (*Virgin of the World*) ascribed to *Hermes Trismegistus* (associated with the Egyptian God *Thoth*), names the four elements as *Fire, Water, Air* and *Earth.*

Although modern science does not support the classical elements as the material basis of the physical world, in the ancient Egypt era,

10. Rochberg, Francesca (December 2002). "A consideration of Babylonian astronomy within the historiography of science" (PDF). Studies in History

11. Philosophy of Science. 33 (4): 661–684. CiteSeerX 10.1.1.574.7121. doi:10.1016/S0039 3681(02)000225. Archived from the original (PDF) on 20101229. Retrieved 20171024

12. Gopal, Madan (1990). K.S. Gautam (ed.). India through the ages. Publication Division, Ministry of Information and Broadcasting, Government of India. p. 78.

four and five could have been a commonality between the number of pyramid faces and the number of classical elements. This classification means the material world in ancient times might be why Egyptian designers used pyramids. But what about number eight?

In our modern digital technology world, we use the binary number system. A byte is a unit of digital information that most commonly consists of eight bits. The largest single-digit number in the binary numeral system is eight (2^1=2, 2^2=4, 2^3=**8**, 2^4=16, etc.).

But, how about the numbers in ancient times? I searched more to see if the numbers four, five, and eight had any other aspects to ancient Egyptians. The Egyptians invented the first ciphered numeral system.

Eight is the first number, neither prime nor semi-prime, and eight is the smallest cubic number: 2^3 =8.

Eight is also considered a lucky number, especially in Japan. This belief is due to its shape –八– called **Suehirogari** (末広がり), which is a

Japanese term that implies prosperity. It is a lucky number because it widens at the bottom, which reminds one of prosperity and growth.[13]

The tradition of depicting humans who have become immortals is an ancient practice in Chinese art. For example, the **8 Immortals** are a group of legendary **xian** (immortals) in Chinese mythology. Each immortal's power can transfer to a vessel (法器) to bestow life or destroy evil.

Together, these eight vessels are called the "**Covert 8 Immortals**." They are said to live on a group of five islands. You can see the correlation of numbers in this Chinese legend as signs of prosperity and longevity.

In Islam, eight is the number of angels carrying the throne of Allah in heaven. Eight is also the number of gates of Heaven. Although Islam came many centuries after the Pyramids era, there might exist a correlation with ancient Egyptians because they were among the first civilizations to believe in an afterlife.

---

13. https://www.lingualift.com/blog/luckyunluckynumbersjapan/

Also, in Arabic, Urdu and Farsi (Persian language), the number eight is written as – – called **Hasht** (YXW), which seems similar to *Japanese Suehirogari*. Perhaps, ancient Egyptians had the same philosophy to build pyramids because it has a broad base and a sharp top.

So, what do all these means? Perhaps pyramids have eight edges to represent luck. But four is considered an unlucky number in Japan. Pronounced "*yon*" or "*shi*," which has the same pronunciation with (死) that means "*death*" or "*die*." Recently, many hotels and hospitals tend to avoid having rooms or floors with the number four. Perhaps that is why Egyptians used square-based pyramids on their first floor. The square area in the base of a pyramid becomes smaller as it ascends to the upper floor. The Bottom always has the maximum extent, and the area becomes zero at the highest level.

We know pyramids were also for religious faiths. Many Egyptian pyramids are designed and constructed for the death of the Pharaohs. Archaeologists (the **Australian Museum**) have found funerary text inscriptions inside pyramid chambers built between 2,375 and 2,160 BCE, which serve the sole purpose of instructing the dead Pharaoh's soul how to cross over to the afterworld. Thus, a Pyramid can represent the form of the physical body (with 4 elements in its square-based), emerging from the Earth and ascending towards the light and eternity at its tip (fifth element). There are '*Pyramid Texts*' found written on the walls of chambers inside the Old Kingdom pyramids. Some texts became known as the '*Coffin Texts*' because mostly they were written on coffins. At the start of the New Kingdom (about 1,500 BCE), a funerary text was made available to Egypt's general population. This text is known today as the '**Book of the Dead**.' [14.]

From another perspective, sometimes, in numerology, the digits are added together; the numbers one through nine are often added together double-digit numbers to make a single-digit number (a process called reducing). For example, 25 becomes 7 (2+5=7). Let me show some exciting outcomes.

---

14. https://australianmuseum.net.au/learn/cultures/internationalcollection/ancientegyptian/ funerarytexts-inancientegypt/

- When adding the numbers 4, 5, and 8 with each other, the sum becomes 17. Then again, when adding 1 and 7 together, the sum is 8.

- The first three digits of the π number are 3.14. When adding the numbers 3, 1, and 4 with each other, the sum becomes 8.

- The first three digits of the **Golden Ratio** number is 1.61. When adding the numbers 1, 6, and 1, the sum also becomes 8.

A recent study shows Giza's Great Pyramid base, not exactly a square, and the shape is not a four-sided figure. British Air Force pilot *P. Groves* in 1940 showed each of the Pyramid's four sides are evenly split from base to tip by very subtle concave indentations and make it an eight-sided figure. This concavity divides each of the apparent four sides in half, creating an incredibly special and unusual eight-sided pyramid, and it is executed to such an extraordinary degree of precision as to enter the realm of the uncanny.

Some believe these subtle lines are only visible from above and at dawn and dusk on the spring and fall equinoxes. This incident leads some conspiracy theorists to think that ancient Egyptians built the pyramids to, perhaps, communicate with something above. The Great Pyramid of Giza has eight faces with eight edges, and its base has eight edges.[15]

Pyramids always remained a mystery to man. Much to surprise, we realize that there is more to uncover with every discovery than ever before!

---

15. https://curiosity.com/topics/theeightfacesofthegreatpyramidofgizacuriosity/)

With mathematics, geometry, philosophy and numerology, the reason for building pyramids versus other shapes of structures in ancient Egypt is much clearer. But these locations still require additional study to find the truth.

The next section is a study of another exciting site in a larger territory.

Part 8
Church of Saint George

This section's study is about the Church of Saint George in northern *Ethiopia* that was carved downwards from volcanic tuff. This substance is the sole architectural material that was in the structure. The Church of Saint George has been dated to the late 12th or early 13th century A.D. and thought to have been during

King Gebre Mesqel Lalibela's reign of the late *Zagwe* dynasty. [16]

The Church of Saint George is among the best known and last built of the eleven churches in the *Lalibela* area. People refer to it as the *"8th Wonder of the World."* [17]

The next table presents some information about the Church of Saint George.

Group	Name	Approximate Build Time	Longitude (Degrees)	Latitude (Degrees)	Latitude (rounded)
1	Church of Saint George	late 12th or early 13th century AD	39.041145°	12.031625°	12.0°

Table 10–Church of Saint George information

16. Moriarty, Colm. "St. George's Church, Ethiopia". Irish Archaeology. Retrieved 29 May 2015
17. "Lalibela: The Eighth Wonder of the World". Tzu Chi Foundation. Archived from the original on 21 January 2013. Retrieved 10 November 2006.

Note that the nearest tenth place of Latitude and Longitude ends with 0 after rounded, which puts it in Group 1.

The next Figure shows the Church of Saint George in **Lalibela**, home to eleven spectacular churches carved inside and out from a single rock some 900 years ago.

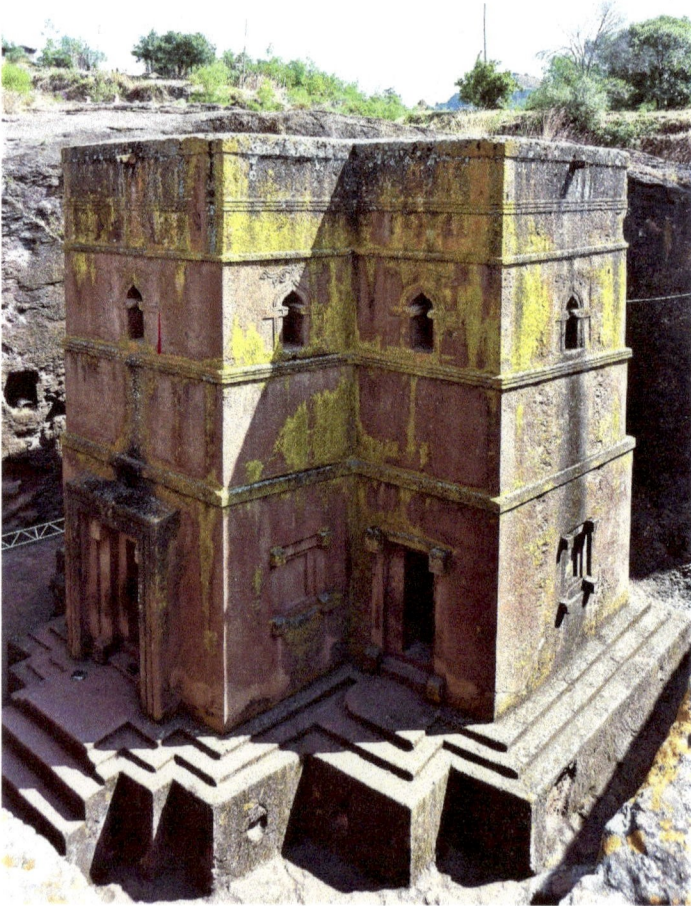

Figure 29–Church of Saint George
https://commons.wikimedia.org/wiki/File:Bete_Giyorgis_03.jpg
Attribution: Bernard Gagnon [CC BYSA 3.0
(https://creativecommons.org/ licenses/bysa/3.0)]

You might wonder if the Ethiopians got the idea of their structures from the Persians who built the Kaaba Zartosht (Ka'baye Zartosht about 550–330 BCE) discussed in Part 4 of this article. You can compare both structures' figures and see their similarities, such as their size, basements, surrounding walls, and entrance stairs.

The next Figure is by Google Earth that shows the Church of Saint George from above and north wise. It shows that the structure is almost aligned with the North-South direction and perhaps with a slight clockwise turn towards the East. The yellow line in the Figure is from the center of this structure that indicates the geographic north direction for easier understanding.

The green line is from the Giza Pyramids, with about 23.92° from the North.

The red line is from the center of Kaaba, with about 4.70° from the North. This 4.70° difference is minor compared to the long-distance to Kaaba. These alignments might make you think of a miscalculation.

Note that the angle between the red line and the green line is about 28.62°, close to the popular 30° angle. This number is vital in geometry. Again, the location and the direction of the structure might not have been chosen and designed randomly. And there seems to be a tendency towards the sites of Kaaba and Giza Pyramids.

Figure 30–Church of Saint George Directions
The green, yellow, and red lines are added.
Google Earth Figure date 3/16/2018, eye alt: 8362 ft. from above,
North wise.
IMAGE: CNES / Airbus

Somehow, the **Ethiopian** engineers might have been aware of Kaaba and Giza Pyramids' geographic location to build this massive structure in this direction, shape and location.

By reading this article up to here, you might have become more interested in studying the old constructions with the perspective of their geographic coordination system, their geometric calculations, and their awareness of other structures in the world. But what about other places?

To continue, I will show what some European engineers did.

Part 9
Arc de Triomphe

Since the 17th century, **Paris** has been one of Europe's major financial centers, diplomacy, commerce, fashion, science and the arts. The city of Paris is the center and seat of the French government. The ***Arc de Triomphe de l'Étoile*** and the ***Arc de Triomphe du Carrousel*** are among the most famous monuments.

Their construction began in 1806.

Figure 31–Arc de Triomphe de l'Étoile
Attribution: Vassil [Public domain]

The Arc de Triomphe de l'Étoile, at 48.873785°, 2.295152°, was the largest arch of its kind in that time. It stands at the center of the **Place Charles de Gaulle**. It is considered the linchpin of the historic axis (**L'Axe Historique**), a sequence of monuments and grand thoroughfares on a route that stretches from the **Louver Palace's** courtyard to the outskirts of Paris. [18]

18. Raymond, Gino (30 October 2008). Historical dictionary of France. Scarecrow Press. p. 9. ISBN 9780-810850958. Retrieved 28 July 2011

Figure 32–Arc de Triomphe du Carrousel
Attribution: Thesupermat / CC BYSA

The Arc de Triomphe du Carrousel, at 48.861750°, 2.332887°, is a triumphal arch masterpiece with many intricate details and decorative additions in Paris in the ***Place du Carrousel***. It is an example of Corinthian style architecture, which is the last developed of the three principal classical orders of ancient Greek and Roman architecture. It was built between 1806 and 1808 to commemorate Napoleon's military victories of the previous years.[19]

The next table represents each monument's information, including their rounded latitude to the nearest tenth place.

---

19   ·http://famouswonders.com/arcdetriompheducarrousel/

Group	Name	Approximate Build Time	Latitude (Degrees)	Latitude (rounded)
1	Arc Du Tiomphe de l'Étoile	1806 Started	48.873785°	48.9°
	Arc de Triomphe du Carrousel		29.975997°	48.9°

Table 11–Arc de Triomphe de l'Étoile and Arc de Triomphe du Carrousel

Note that their rounded latitude of 48.9° is in Group 1, with 9 in their nearest tenth place. You might wonder why.

Monuments Facing Kaaba

The next Figure is by Google Earth from above, and North wise shows a portion of Paris. In the top-left of the Figure is the Arc de Triomphe de l'Étoile monument (Spot A). The avenues radiated from the Arc de Triomphe monument represent a precise geometric design, which does not seem randomly built. The straight yellow line is stretched on the main street from Arc de Triomphe to Arc de Triomphe du Carrousel (Spot B), which is about 115.91° from the north. You might wonder why this street direction was necessary.

Since both monuments' face is almost towards Kaaba, it got my attention to hunt more into it. The red line in the Figure is from Arc de Triomphe de l'Étoile monument (Spot A) to Kaaba, which is about 119.12° from the north. The angular difference between these two red and yellow lines is only 3.21°. In other words, if Spot B were built about 175 meters to the South and on the red line, there would have been no angular differences.

Figure 33–Arc de Triomphe de l'Étoile and the Arc de Triomphe du
Carrousel Red and yellow lines and the letters are added.
Google Figure date 2/25/2019, eye alt: 9437 ft. above and North wise.
IMAGE: Maxar Technologies

Nevertheless, this 3.21° difference is small. Due to the long distance to Kaaba and the excellent geometric designs of the city streets, it seems like another tiny miscalculation similar to the Taqe Bostan road. By comparing this street with the Taq-e Bostan Boulevard, discussed in Part 3, you can see how both roads are about 3° to 4° off from Kaaba.

We all understand there are many streets in Paris and other parts of the world that are elongated towards Kaaba accidentally.

However, the similarities make me wonder if there is a pattern.

Remember, my study is solely to find clues.

In the next section, I want to check Asia areas and see any sign of interest in their constructions towards Kaaba and Giza Pyramid's locations.

Part 10
Leshan Giant Buddha

According to consensus in modern genetics, anatomically modern humans first arrived on the Indian subcontinent from Africa between 73,000 and 55,000 years ago. However, the earliest known human remains in South Asia date to 30,000 years ago. Settled life, which involves the transition from foraging to farming and pastoralism, began in South Asia around 7,000 BCE.[20, 21]

Siddhartha Gautama, known as *Buddha*, was a monk (*rama a*), mendicant, sage, philosopher, teacher and religious leader teaching Buddhism. Many believe him to have lived and taught mostly in the ortheastern part of ancient India sometime between the sixth and fourth centuries BCE.[22, 23, 24]

20. Michael D. Petraglia; Bridget Allchin (2007). "The Evolution and History of Human Populations in South Asia: Interdisciplinary Studies in Archaeology, Biological Anthropology, Linguistics and Genetics. Springer Science & Business Media." p. 6. ISBN 978140205562
21. Boeree, C George. "An Introduction to Buddhism". Shippensburg University. Retrieved 10 September 2011.
22. Warder, AK (2000), "Indian Buddhism, " Delhi: Motilal Banarsidass
23. Dhirasekera, Jotiya. "Buddhist monastic discipline." Buddhist Cultural Centre, 2007
24. Shults, Brett. "A note on Sramana, in Vedic Texts." Journal of the Oxford Center for Buddhist Studies 10 (2016)

A structure that interests me in this research is the Leshan Giant Buddha. This statue is 233 ft. (71 meters) tall, stone-built between 713 and 803. It is at 29.544722°, 103.773333° located in **Sichuan**, People's Republic of China. The leader of construction was a Chinese monk named **Hai Tong**.

He hoped Buddha would calm the turbulent waters that plagued the shipping vessels traveling down the river in front of it.[25]

The Leshan Giant Buddha is about 2,031.82 miles away from the Equator.

Figure 34–Leshan Giant Buddha
https://upload.wikimedia.org/wikipedia/commons/3/38/Leshan_Buddha_Statue_View.JPG
Attribution: Ariel Steiner [CC BYSA 2.5
(https://creativecommons.org/licenses/bysa/2.5)]

---

25. Bryan Hill. "The Leshan Giant Buddha: Largest Stone Buddha in the World", 22 July 2018 Ancient Origins

Group	Name	Approximate Build Time	Latitude (Degrees)	Latitude (rounded)
1	Leshan Giant Buddha	between 713 and 803	29.544.722°	29.5°

Table 12–The Leshan Giant Buddha in Sichuan, China.

Note that the nearest tenth place of its rounded latitude is five, not in the discussed range of any groups in Part 1; perhaps it can be in a new group named Group 3.

The next Figure is from the side for easier understanding of its directions. I added the 3D building and the yellow-, blue-, and teal-colored lines.

Figure 35–Leshan Giant BuddhaDirections
Colored lines are added
Google Earth Figure date: 1/30/2018, eye alt: 1800 ft., almost West wise.
IMAGE: Maxar Technologies

Another impressive fact is the 29.5° latitude of the Leshan Giant Buddha, which is remarkably close to the latitude number of the remains of Persepolis with 29.9°, and the latitudes number of the three Giza Pyramids and the Sphinx with 30°. More precisely, this is only 0.43° difference of Khufu Pyramids latitude, which means 29.54 miles. Note that the number 29.54 is almost 30, which is a popular number in geometry.

In other words, if the designers built the Leshan Giant Buddha only 29.54 miles to the North, it would have had the same latitude as the Khufu Pyramid in Egypt. You might want to know if all of this numeral correlation is accidental.

It is not clear why the giant statue that caused the extreme effort to build is facing Giza Pyramids. It might be a nice coincidence. Again, you might think there are many statues in the world, each facing towards something, but the numbers here make you wonder of an intention.

Part 11
Christ the Redeemer Statue

Let us have a peek at another gigantic monument from the other side of the world. The Christ the Redeemer statue in *Rio de Janeiro*, Brazil, is located at -22.951944°, -43.210556°. It is a symbol of Christianity globally; the art has also become a cultural icon of both Rio de Janeiro and Brazil. It is one of the "*New 7 Wonders of the World.*"[26]

The next Figure shows the Christ the Redeemer statue from behind at the top of a mountain in Rio de Janeiro, Brazil, facing the Ocean.

---

26. The Christ the Redeemer Statue Rio de Janeiro, Brazil .... https://www.worldatlas.com/articles/the-christtheredeemerstatueriodejaneirobrazil.html

Figure 36–Christ the Redeemer statue
https://commons.wikimedia.org/wiki/File:Morro_do_Corcovado.jpg
Attribution:
Ulysses Rj [CC BYSA 3.0 (https://creativecommons.org/licenses/bysa/3.0)]

Group	Name	Approximate Build Time	Latitude (Degrees)	Latitude (rounded)
1	Christ the Redeemer	between 1922 and 1931	-22.951944°	-23.0°

Table 13–Christ the Redeemer statue in Rio de Janeiro, Brazil.

Table 13 showcases some of the Christ the Redeemer statue information. Note that the rounded latitude of -23.0°, with "0" in its nearest tenth place, is in Group 1. Is this another coincidence?

This statue faces approximately 83° from the North, about 7° from the East. As a reminder, the Leshan Giant Buddha represented in the previous section, faces about 7.90° from the West.

The direction of the Christ the Redeemer statue towards the Leshan Giant Buddha is 70.31°, 85.27° from the North in 10,354 miles.

- The Christ the Redeemer statue faces Kaaba by about 67.63° from the North. This number is 7.63°, more than 60°.
- Moreover, the statue's trend towards Giza Pyramids is
- approximately 56.54° from the North, which is 3.46° less than 60°.
- Christ the Redeemer statue is facing the center of **Sha'Z Square**. The angle is 59.42° from the North, which is 0.58° less than 60°.

Again, we see the use of the number 60, which is popular in geometry.

Thus, you saw how these gigantic human-made monuments, the 'Christ the Redeemer,' the 'Leshan Giant Buddha,' and the 'Colossi of Memnon' statues from a different side of the globe have a common denominator.

We only have a little glimmer of what is going on. The designers could build their statues in any direction. How was this done?

In the next section, I will study a structure in a different environment.

Instead of searching on land, I am going to explore underwater.

Part 12
Atlantic Anomaly Rectangle

It is an old saying, "The Ocean of the world hides a lot of history and a lot of mystery." Earth's oceans cover about 71 percent of the planet. The Atlantic Ocean's spread is from the North Pole to the South Pole and between African, European, and American continents.

The next Figure shows an underwater area from above and North wise.

There is an unfamiliar model on its floor at about 31.40°, -24.00°.

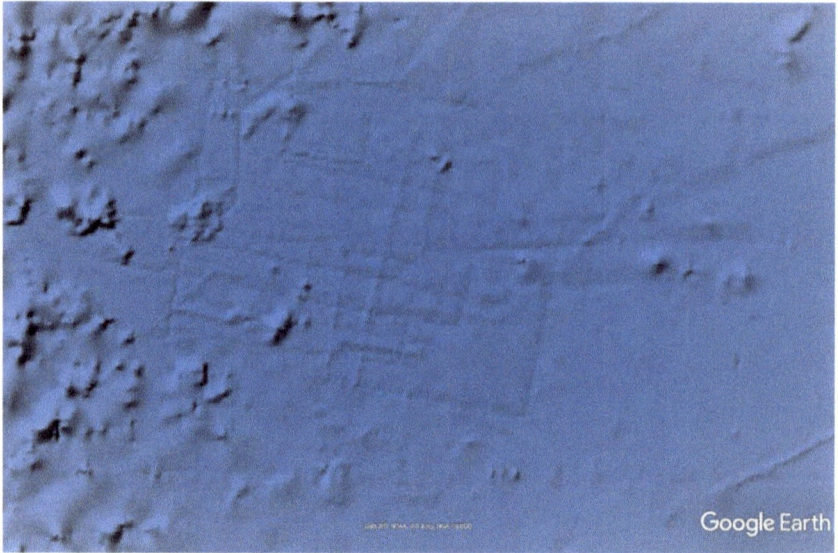

Figure 37–Atlantic Anomaly Rectangle
Google Earth Figure date: 12/13/2015, eye alt: 200.61 mi, north wise
IMAGE: Landsat / Copernicus

Looking carefully at the last Figure, I speculated there was once a rectangular structure there, like the ruins on the ground with right angles shown on some of the earlier figures of this article. Perhaps this remnant structure was submerged under the ocean in time. Check the long and straight lines in the middle area of this Figure. You can see it with Google Earth as well. I am eager to analyze this anomaly because it can be intriguing under the water evidence. I named it the "*Atlantic Anomaly Rectangle*."

- The next Figure shows the same last picture with additional information. The Red Line is an indicator in exactly 100 miles, with 100° from the North. It is about 10° from the East-West.

- The White Rectangular shape of ABCDA shows the borders.

- The corners seem to be right-angled, which is strange in that natural region.

- Dimensions are approximately 100 x 75 miles, and the area is about 7,800 miles².

- The small Yellow Circle is Spot M, at 31.40°, -24.00° as another indicator. This spot is in Group 2.

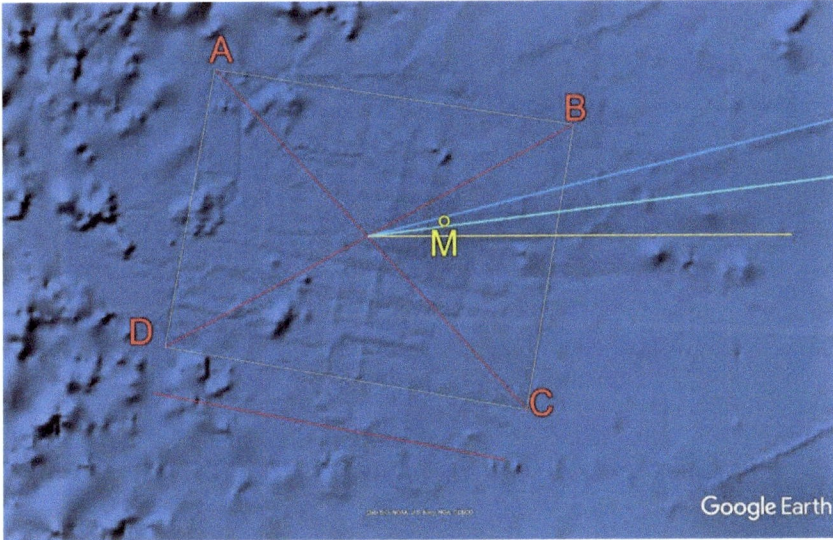

Figure 38–Atlantic Anomaly Rectangle
Same as Figure 37, but colored lines and letters are added

Note that Spot M is about 21.63 miles away from the center of the Atlantic Anomaly Rectangle area. It is approximately 78.48° from the North. Due to the *"Seafloor Spreading,"* the white rectangle area is moving along the time, and its center today is about 600 miles away from the nearest **Atlantic ridge**. The "Seafloor Spreading" is a geologic process in which tectonic plates—large slabs of Earth's lithosphere— split apart from each other. This incident separates the **North American plate** from the **Eurasian Plate** and the South American Plate from the African Plate. It is unclear if Spot M and the center of the Atlantic Anomaly Rectangle were over each other in a time.

The yellow line is over the straight anomaly line on the floor of this place. Refer to the previous Figure as a source to see the line. Interestingly, the line is 90° from the North.

The blue line is towards Giza Pyramids with 76.50° from the North, which is only 6.50° more than 70°. The teal line is towards Kaaba with 83.13° from the North, which is 6.87° shorter than the popular 90°. Both 6.50° and 6.87° seem similar to each other and become 7° when rounded.

Interestingly, the teal line is in the middle of both yellow and blue stripes. The teal line is about 6.87° and 6.63° angle respectfully from each. Again, both of these numbers are similar and become 7° when rounded. Making you wonder deeper about the geometric correlation.

Note how the position of Spot M is between the lines that are towards Kaaba and Giza Pyramids.

As mentioned, **Sha'Z Square's** center was named "**CC3**" with a latitude of 32.632275°. By comparing it to the center of this Atlantic Anomaly Rectangle with 31.343306°, you can see they only have about 1° difference (exactly 1.288969°). This minor difference means both are almost in the same latitude from the Earth's equator.

The next Figure shows a Google Earth Figure of **Sha'Z Square** (right side of the Figure) and the Atlantic Anomaly Rectangle (left side of the Figure).

Figure 39–Atlantic Anomaly Rectangle and Sha'Z Square Area Colored
lines are added.
Google Earth Figure date: 12/2016, eye alt: 3,058.72 mi, north wise
IMAGE: Landsat / Copernicus

The yellow lines are the Geographic Coordinate System Grids to compare the previous paragraph's latitudes, more comfortable.

The straight brown line connects the centers of both **Sha'Z Square** and the Atlantic Anomaly Rectangle.

Interestingly, the brown line almost overlaps the diameters of each area. You might want to know if the Atlantic Anomaly Rectangle was, by nature, a map error or our ancestors built.

You can stretch lines to other places such as the Spot P, Persepolis, and Kaaba Zartosht Spots to study further. Refer to Figures 12, 15, and 16.

You can also compare the shape of the Atlantic Anomaly Rectangle with the other sites in Figure 13 to find more surprises.

The next Figure shows **Sha'Z Square** (top-right), the Atlantic Anomaly Rectangle (top-left), and the Church of Saint George (downright).

Figure 40–Atlantic Anomaly, Sha'Z Square, and Church of Saint George
The lines and the names are added.
Google Earth Figure date: 12/30/2016, eye alt: 2,935.91 mi, north wise
IMAGE: Landsat / Copernicus

The elongation of the Atlantic Anomaly Rectangle is almost towards the Church of Saint George with the purple line. The distance is 4,228.71 miles, and it is 93.26° from the North. This number is only 3.26° more than a 90° angle from the North. Suppose they built the Atlantic Anomaly Rectangle was intentionally towards the Church of Saint George location. In that case, having about 3° difference in about 4,000 miles is a minor miscalculation due to their technology.

The three lines on the map represent a large triangle shape. This triangle is vaster than the one specified in Part 5 of this article. However, people from these three spots cannot see each other's sites from the Earth's surface due to the global curve. So, you might wonder what the purpose of creating such a vast triangle shape is except for navigation.

If they altered the Atlantic Anomaly Rectangle's location, its diameter and elongation would not have reached *Sha'Z Square* and the Church of Saint George. You might think those designers made a wise decision in the direction of those days.

86

The next Figure shows a closer look at the **Sha'Z Square** area. The green line is from Taq-e Bostan (Spot D) to the Church of Saint George. The Spot CS is the intersection of the green line and the diameter A to C of **Sha'Z Square**. Surprisingly, **Sha'Z Square**'s diameters are precisely towards the Atlantic Anomaly Rectangle to the West and almost towards the Church of Saint George to the South.

Figure 41–Sha'Z Square area Directions
Lines, circle, and names are added.
Google Earth Figure date: 12/30/2016, eye alt: 416.54 mi, north wise.
IMAGE: Landsat / Copernicus

The Spot CS is almost in the middle of the diameter of A to C. To show this matter better, I drew a green circle, centered in CS and radiused to A.

You can see the distances from CS to D or B are similar. In other words, CS has an equal distance to D and B as to A and C.

The green line from Taq-e Bostan (Spot D), located at the Northern corner of **Sha'Z Square**, to the Church of Saint George, is 1,621.17

miles long with 200.15° from the North, which is only 5.74° difference from the diameter of D to B.

The brown line is from CC3 (the intersection of the diameters) to the Atlantic Anomaly Rectangle center. It almost overlaps with the red diameter A to C, discussed earlier.

Overall, we can see how the triangle shape of Taq-e Bostan, the Church of Saint George, and the Atlantic Anomaly Rectangle in Figure 40 is close to a Right-Angled Triangle.

Amazingly, the Church of Saint George with the Atlantic Anomaly Rectangle centers and **Sha'Z Square** creates a particular Right-Angle Triangle with 89.27° at the center of **Sha'Z Square** corner. This number is only 0.73° less than the popular 90°. The angles are as follow:

- Center of the Church of Saint George corner: 78.06°
- Center of the Atlantic Anomaly Rectangle corner: 25.00°
- Center of **Sha'Z Square** corner: 89.27°

This Right-Angled Triangle's design in the vast area of about 9,810 $mile^2$ makes it hard to believe in coincidence. It is impossible to see it from the Earth's surface, and it is only visible from the sky. You might wonder if the Persian architects knew about all these calculations but somehow made less than a degree mistake.

You might also think their calculations were precise at the time, but the Earth had shifted its lands in time that changed today's spot coordination.

Huge Isosceles Triangle

Another shocking discovery is the measure from Spot M of the Atlantic Anomaly Rectangle to Kaaba and Christ the Redeemer statue. The distances are as follows:

- Spot M to Kaaba: 3,962.94 miles
- Spot M to Christ the Redeemer statue: 3,946.11 miles

The difference is only 16.83 miles. This number is almost nothing within 4,000 miles. This equivalent distance is an Isosceles Triangle, much larger than the Taq-e Bostan Triangle in Part 5. The angles are as follow:

- **Spot M** corner: 83.59° almost a Right Angle
- **Kaaba** corner: 48.24°
- **Christ the Redeemer statue** corner: 48.17°

Breathtakingly, by minor differences, the Spot M corner is almost a right angle, and the other two angles are equal. Thus, this huge triangle is a Right-Isosceles Triangle with two identical lengths and angles and one right angle. Again, one might ask if this is another coincidence.

Geographic North Pole

The North Pole, also known as the **Geographic North Pole** or **Terrestrial North Pole**, is the point in the **Northern Hemisphere** where the Earth's axis of rotation meets its surface. It is the Earth's northernmost point.

The distance from Spot M in the Atlantic Anomaly Rectangle to the Geographic North Pole is 4,056.27 miles. This number is the same as to Taq-e Bostan site with 4,053.39 miles. There is only less than three miles difference in the distance of more than 4,000 miles.

Spot M's equidistance with the North Pole and Taq-e Bostan is another astonishing discovery about how all these spots correlate with each other.

You might have questions such as:

- Is the Atlantic Anomaly Rectangle deep under the ocean, an ancient human-made structure?

- Were the **Sha'Z Square** builders aware of the Atlantic Anomaly Rectangle?

- Is the Atlantic Anomaly Rectangle associated with the lost city of **Atlantis**?

- In continuation of my research, I will go back to the European area, but this time within an older era, to search for more shreds of evidence.

Part 13
Tumulus of Bougon

The Tumulus of Bougon is at **Bougon** near **La-Mothe-Saint-Héray**, somewhere between **Exoudon** and **Pamproux** in **Nouvelle-Aquitaine**, **France**. It is a group of five prehistoric monuments of **Neolithic** barrows. Tumulus of Bougon is an old complex of tombs with varying dates, which its oldest structure in about 4,800 BCE. The 1,000-year development of architecture at Bougon can be in three phases: [27]

1. Round or oval mounds with corbel-vaulted chambers
2. Elongated mounds with small rectangular megalithic chambers
3. Large rectangular megalithic chambers

The shapes of the architectures show very well their interest in basic geometry. It convinced me to study their location and designs deeper.

The next table represents some information, including the geographic coordinates of the site.

---

27. https://en.wikipedia.org/wiki/Tumulus_of_Bougon

Group	Name	Approximate Build Time	Longitude (Degrees)	Latitude (Degrees)	Latitude (rounded)
2	Tumulus of Bougon	4,800 BCE	0.066660°	46.373795°	46.4°

Table 14–Tumulus of Bougon information

Note how the rounded latitude to its tenth place is 46.4°, which puts it in Group 2. Compared to other locations studied earlier in this article, these coordinates tell me this spot was probably not chosen by accident.

The next Figure shows the site by Google Earth. The Tumulus of Bougon site is on the left side of the Figure. The teal color line is to Kaaba with 2,819 miles and 1 from the North. The dark blue line (close to the teal line) is to the Giza Pyramids with 2,020 miles and 113.13° from North. Yes! The angles are almost identical.

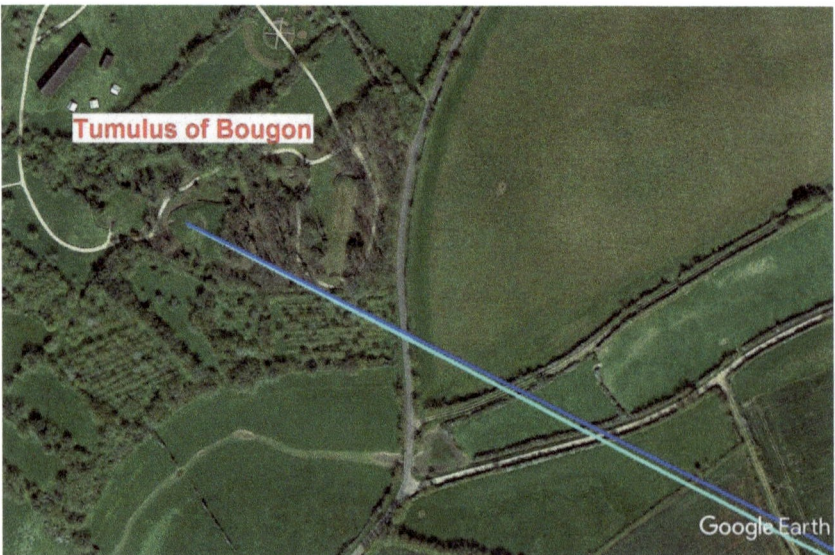

Figure 42–Tumulus of Bougon site
Light and dark blue lines and the name in red are added.
Google Earth Figure date: 4/11/2015, elevation: 300ft, eye alt: 1926ft

There is an 801 miles difference between Kaaba and Giza Pyramids, which is almost a round number of 800 miles. Also, there is an angular difference of only 0.8°, which is minor due to the long distances. Note how both numbers of 800 and 0.8, which are their differences in length and angle, are alike, and both use #8. In other words, the location of the

Tumulus of Bougon site is almost on the line of Kaaba to Giza Pyramids. The question is: "Were designers aware of the Kaaba and Giza Pyramids locations in that era?"

Part 14
Tumulus of St. Michel

About 165 miles North-West of the Tumulus of Bougon is the **Tumulus of St. Michel**. The Tumulus of St. Michel is a megalithic grave mound East of **Carnac** in **Brittany**, **France**. It is said to be the largest of its kind in Europe, built during the fifth millennium BC.

The next table represents some information, including the geographic coordination of the site.

Group	Name	Approximate Build Time	Longitude (Degrees)	Latitude (Degrees)	Latitude (rounded)
3	Tumulus of St. Michel	4,500 BCE	—3.073689°	47.588244°	47.6°

Table 15–Tumulus of St. Michel information

The rounded latitude to its tenth place is 47.6°, which does not put it in any of the Groups, but I can put it in Group 4, similar to the **Leshan Giant Buddha** statue in Part 10.

The next Figure shows the area. However, same as the Tumulus of Bougon, this site is almost aligned with the Kaaba and Giza Pyramids.

The Tumulus of St. Michel is at the top-left of the Figure. The teal color line is to Kaaba with 2,982.75 miles and 112.09° to the North. The dark blue line is to Giza Pyramids with 2,183.22 miles and 111.51° from the North. The angles are almost identical.

Also, there is an angular difference of only 0.51°. Half a degree is minor due to the long distances they have. In other words, the location of the Tumulus of St. Michel site is almost in the line of Kaaba to Giza Pyramids. The length shows a difference of 800.47 miles between Kaaba and Giza Pyramids, which is again a round number of 800 miles. This correlation is like the Tumulus of Bougon site and seems incidental.

Figure 43–Tumulus of St. Michel site
Light blue and dark blue lines and the name are added
Google Earth Figure date: 7/14/2018, elevation: 84ft, eye alt: 1480ft

Thus, besides all the purposes of these structures that researchers have discovered before, with these discoveries, you might think the builders in France, in that era, were aware of the Kaaba and Giza Pyramids locations, and it was important for them to build their structures in a place aligned with them.

If this is true, then perhaps the Giza Pyramids and Kaaba were made before 4,800 BCE., or the engineers were interested in the spots before. Either way, it seems likely that both the Tumulus of St. Michel and Tumulus of Bougon was built deliberately in the locations. You can compare the elongation of this structure with the ones in Figure 13. This section needs further investigation.

In the next section, I want to study an older place and much further to the Giza Pyramids and Kaaba area. It is in the Southern American continent.

Part 15
Ancient Ruins of Caral

The ***Sacred City of Caral*** is a 5,000-year-old metropolis representing the Americas' oldest known civilization, known as the ***Norte Chico***. It is in the ***Supe Valley***, near ***Supe***, ***Barranca Province***, ***Peru***. Caral is the largest recorded site in the ***Andean*** region. Caral may answer questions about the origins of the ***Andean civilizations*** and the development of the first cities.

The next table shows some information about three of the pyramids in the ruins of Caral. Location is -10.890498°, -77.521502°.[28]

Group	Name	Approximate Build Time	Latitude (Degrees)	Latitude (rounded)
1	The Gallery	4,000 - 4,600 BCE	-10.893484°	-10.9°
	La Huaca		-10.893573°	-10.9°
	Pyramid of Lesser		10.891838°	-10.9°

Table 16–Ruins of Caral Information

Note that each rounded latitude to their nearest tenth place is "-10.9°" that is in the range of Group 1.

28. https://en.wikipedia.org/wiki/Caral

By comparing the numbers with the Giza Pyramids' latitudes in Part 7 of this article, you might wonder if there is even more correlation in between.

The next Figure is a view of the site by Google Earth. I added the red squares, lines, and letters to the main Figure.

The red squares are the approximate borders of each three pyramids. The **Lesser Pyramid** is with the letter "L," **La Huaca Pyramid** is "H," and the **Gallery Pyramid** is "G."

Interestingly, you can see some circular structures in the Figure. Usually, there are rectangular forms with right angles in ancient structures.

And, circular designs are rare, especially in that era.

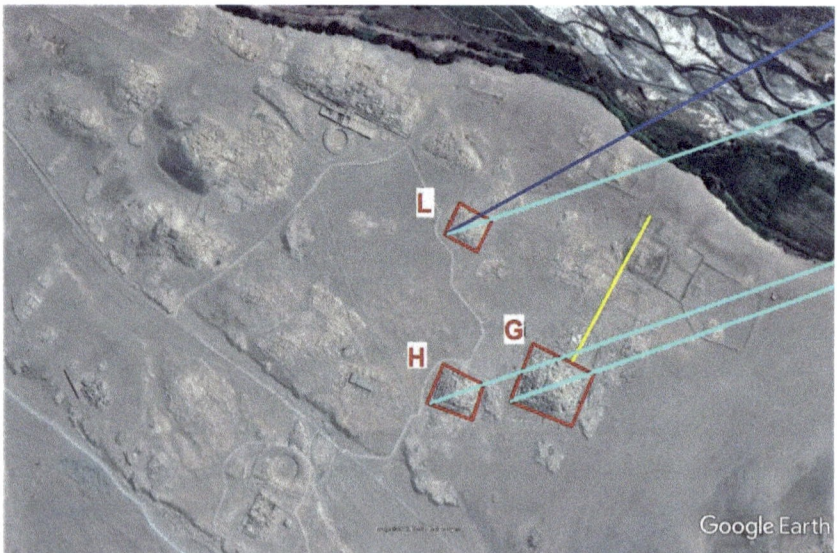

Figure 44–Ruins of Caral
Colored lines, red squares, and letters are added.
Google Earth Figure date: 4/21/2017, eye at: 4047 ft from above, north wise.
IMAGE: CNFS / Airbus

The Pyramids were built about 25° from the North, which seems irrelevant. As a reminder, the Giza Pyramids were exactly north wise.

You might wonder why and is there any common denominator. Their shapes are amazingly almost like **Sha'Z Square** shape, and some other structures showcased earlier. Refer to Figures 13 and 17.

The dark blue line comes from the Giza Pyramids, with 7,708 miles distance and 61.90° from the North. This number has only 1.90° more than the popular 60° in triangles.

The three teal blue color lines are stretched from Kaaba to the Southwest corner of every three pyramids' bases, with approximately 8,247 miles and 71.43° from the North. This number is only 1.43° more than 70°.

By the way, the angle between the two lines from Kaaba and Giza Pyramids that approach the **Lesser Pyramid**, assigned with the letter "L," have about 9.57°. This number is 10° when rounded.

The yellow line on the Figure is over a long-drawn road from the **Gallery Pyramid** and has exactly 30° angle from the North, a popular angle. This road's direction reminds me of Taq-e Bostan and Paris' streets that I discussed earlier how excitingly they also almost face Kaaba and Giza's Pyramids by minor differences.

Note how the teal lines exactly cross the diameters of the pyramid bases. It is similar to the Giza Pyramids in Egypt described earlier in Part

7. Amazing commonality!

The teal blue line from Kaaba to Pyramid "G" has about 41.43° with the yellow line. This number is 3.57° less than the popular 45° angle in geometry, particularly in triangles.

Precise designs require premium calculations. The engineers who built these pyramids could build the structures in any shape, size and direction. They could have been set more precisely with Giza Pyramids and Kaaba locations amidst only small changes.

You might wonder how our ancestors could have known such sophisticated mathematical calculations to find the exact direction within a far distance. Nevertheless, the precise measures of these distinct structures in about 5,000 BCE do not seem accidental. I need more study for this section also.

So far, it seems like some of our engineer ancestors had a tremendous interest in Giza Pyramids and the Kaaba places. If so, then why? Why were these two locations mattering to the ancient people in different parts and eras of the world? Perhaps the answer is wrapped in a wider panorama of the earth. In the next section, I study these areas even further and from the whole globe.

Part 16
Continents

As mentioned earlier about the Kaaba's origin, some historians doubt that Kaaba is the oldest structure on Earth and an important archaeological and historical spot for humankind. Now, let us study this by going back in time.

Scientists say that the continents of our ever-changing earth began to break apart about 175 million years ago. Before that, it assembled from earlier continental units approximately 335 million years ago. That mechanism is called *Continental Shift*. *Pangaea* was the most recent supercontinent to have existed and the first to be reconstructed by geologists.[29]

The next Figure is an illustration of the *Pangaea Supercontinent*. The name of each area is within the borders.

Suppose you can fill the empty area between Eurasia, Africa, India and Australia. In that case, the entire map will look like a green oval field that Egypt is approximately at the center.

As you can understand by the Figure, both Giza Pyramids and Kaaba locations are near the middle of the green oval field, but the Giza Pyramids plateau is closer to the center.

---

29. Rogers, J.J.W.; Santosh, M. (2004), Continents and Supercontinents, Oxford: Oxford University Press, p. 146, ISBN 9780195165890

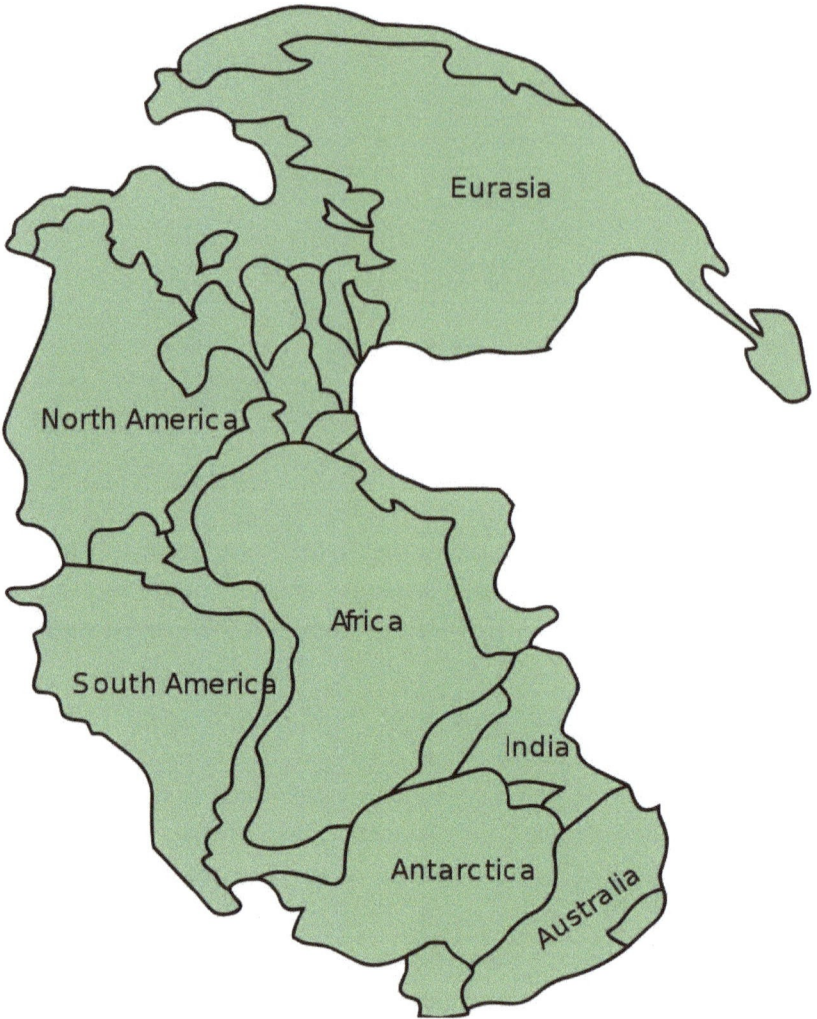

Figure 45–Pangaea Supercontinent
Attribute: Hakim Djendi ~ commonswiki 10/12/2017

Imagine in the same illustration that we could somehow stretch and rotate the Australia, India, Saudi Arabia and Africa sections to attach each other and fill the empty gap in between. In this case, the green area would seem more like a circle than an oval, with the Giza Pyramids,

104

Kaaba, and the **Sha'Z Square** closer to the center. This imagination gave me the idea that perhaps one of these spots is in the lands' center.

So, let me return and continue the study to search for the center of dry land on Earth's surface in our era.

Part 17
The Big Land Circle

Earth has seven continents. They are *Asia*, *Africa*, *North America*, *South America*, *Antarctica*, *Europe* and *Australia*.

By looking at the next Figure, you can see how the world seems exactly from above Kaaba. Entire Africa, whole Europe, and most of the Asia continent is on this side of the planet. You cannot see Antarctica, North America, South America or Australia continents in this Figure.

Figure 46–Earth from above Kaaba
Google Earth Figure date: 12/30/2016, eye alt: 9095.10 mi. North wise.
IMAGE: Landsat / Copernicus

We know the cover of about %71 of Earth's surface is by water, but most of this side of Earth is dry land. Perhaps this Figure shows the best view of the planet covered mostly by land. In reverse, the antipode (opposite side of the globe) would have the most water offered in the next section.

Kaaba's Big Land Circle

By drawing a circle with a radius of 8,773.20 miles (14119.10 kilometers) centered in Kaaba, it will include the entire American continent from North to South. I name this circle the "***Kaaba's Big Land Circle***.". In the next sections, I will show the Spot F, the most distant land at 22.883272°, -109.987259°. Wonderfully, its rounded latitude to the tenth place is "22.9°," which puts it in Group 1. From Kaaba's perspective, this spot is 35.36° from the North, only 5.36° more than the popular 30°.

Kaaba's Antipode

The next Figure is from the opposite side of Kaaba, in the Pacific Ocean, which is called the ***Kaaba's Antipode***.

Figure 47–Earth from above Kaaba's Antelope
Google Earth Figure date: 12/30/2016, eye alt: 9077.58 mi north wise.
IMAGE: Landsat / Copernicus

Kaaba's Antipode is the remotest location of the globe from Kaaba. The geographic coordinates of Kaaba are 21.422500°, 39.826184°, which was discussed earlier in Part 2. By globally calculating the opposite coordinates of Kaaba, you can find its Antipode coordinates at 21.422500°, -140.173816°.

However, using the Google Earth tools, the antipode coordinates are 21.408745°, -140.194917°, different from the calculation one. This twomile difference is because the Earth is flatter at the poles and bulges at the equator.

All the directions from Kaaba's Antipode point are equidistantly facing Qiblah. This attribute might be useful for some tribes or religions such as Islam. Muslims can take any direction for their prayers from that spot because they all face Kaaba at an exact distance.

In the previous Figure, there is some small portion of drylands on

Earth's surface, such as **Eastern Australia**, **New Zealand** (West), **Hawaii** (North), a part of **Antarctica** (South) and a section of the western edge of **America** (East). As expected, most of this side of the planet is water.

Perhaps this is the best view of Earth that is most utmost covered by water.

The next Figure shows the same picture yet including some details. The red spot in the middle of the Figure is Kaaba's Antipode point. The big yellow circle is the outer circumference of "**Kaaba's Big Land Circle**." From this perspective, most lands are outside of this circle and on the other side of the planet. The radius of this circle from the Kaaba's Antipode point is about 3,662.30 miles.

You can see New Zealand in the lower-left, the biggest island outside of "Kaaba's Big Land Circle."

The small red circle in the top right is the same Spot F. Of course, it is also the nearest land spot to the Kaaba's Antipode point, with 3,662.30 miles and 35.39° from the North. This angle is only 5.39° more than the popular 30°. The white area in the Southern part of the Figure is sections of **Antarctica** (South Pole).

You might want to know if Spot F is random?

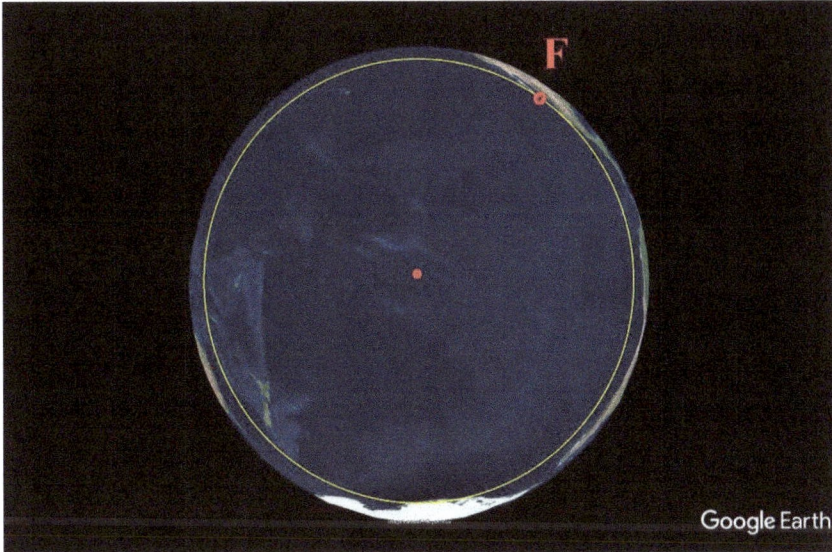

Figure 48–Earth from above Kaaba's Antelope Kaaba's Big Land Circle
Same Figure 46Yellow circle and red dots are added

Azimuthal Equidistant Projection

Maps cannot be without map projections. Projections are a subject of several pure mathematical fields. A ***Map Projection*** is a systematic transformation of the latitudes and longitudes of locations from a sphere's surface or an ellipsoid into places on a plane. The ***Azimuthal Equidistant Projection*** is an azimuthal map projection. It has the useful properties that all points on the map are at proportionally correct distances from the center point and that all points on the map are at the correct azimuth (direction) from the center point.[30]

The next Figure is an ***Azimuthal Map of Kaaba***, a map projection of Earth with the center of Kaaba in a circular shape. Radius: 12,436.74 miles (20,015 km). This map covers the whole world, including the lands (white) and the waters (blue) areas. The round map is split into four inter circular areas by gray circles: the geographic grids.

30. Snyder, J.P. (1989). Album of Map Projections, United States Geological Survey Professional Paper. United States Government Printing Office. 1453.,

I named these circular regions *INSIDE*, *MIDDLE-1*, *MIDDLE2*, and *OUTSIDE* for ease. You can see how the INSIDE area surrounds Kaaba, Saudi Arabia, the Middle East, Europe, and most of Africa and is frequently land (white areas). The OUTSIDE circular area is mostly water (blue regions). Most of Northern America, almost full Southern America, Antarctica, and Australia are inside the MIDDLE-2 round area.

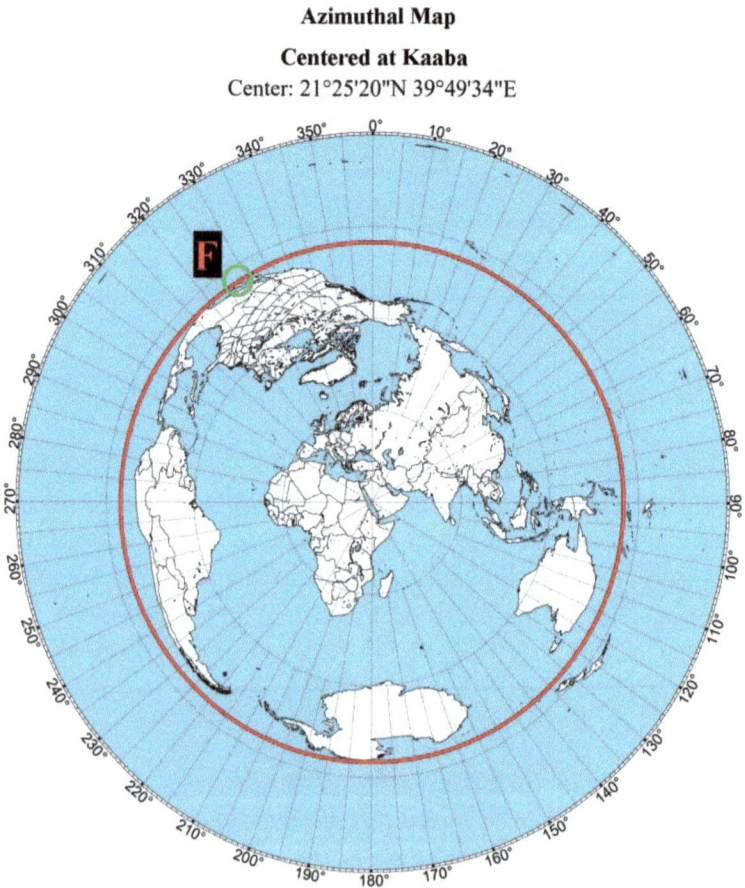

Azimuthal Map

Centered at Kaaba

Center: 21°25'20"N 39°49'34"E

Figure 49–World Azimuthal Map centered in Kaaba, and Spot F Colored circles and letter F are added
Courtesy of Tom (NS6T), https://ns6t.net/azimuth/azimuth.html

In this Figure, I do not know what percentage represents the Water. Still, it would be interesting if it is about 70 percent because water covers about 70 percent of Earth's surface, similar to the human body with approximately 70 percent water. So, what is going on? I will look more later in Part 20.

The Red Circle is "***Kaaba's Big Land Circle***" with a radius of 8,773.20 miles (14,119.10 km) centered at Kaaba.

The Green Circle shows Spot F's location with the approximate geographic coordinate of 22.883272°, -109.987259°.

Note how all the world's major drylands, which cover about 30 percent of Earth's surface, are surrounded by "***Kaaba's Big Land Circle.***" In other words, all the mainland on Earth can fit inside this big red circle area.

Some small islands, such as ***Hawaii*** and ***New Zealand***, which have almost the same distances as Kaaba, are not considered.

Interestingly, by looking at the last Figure, you can see how the four big lands of North America, South America, Australia and Antarctica are almost the opposite of each other, with approximately the same distance from Kaaba. I will discuss this later.

The next Figure is the opposite of the previous Figure. It represents a global azimuthal map projection of the ***Kaaba's Antipode*** spot. You can understand how it holds the lands (white) and the waters (blue) areas.

The Red Circle is "***Kaaba's Big Land Circle,***" and the Green Circle shows the Spot F.

Again, inside this round map is split into four inter circular areas by gray Geographic Grid Circles. I named them ***OUTSIDE***, ***MIDDLE2***, ***MIDDLE-1***, ***INSIDE*** from the Center of the Figure respectfully.

Azimuthal Map

Centered at Kaaba's Antonode

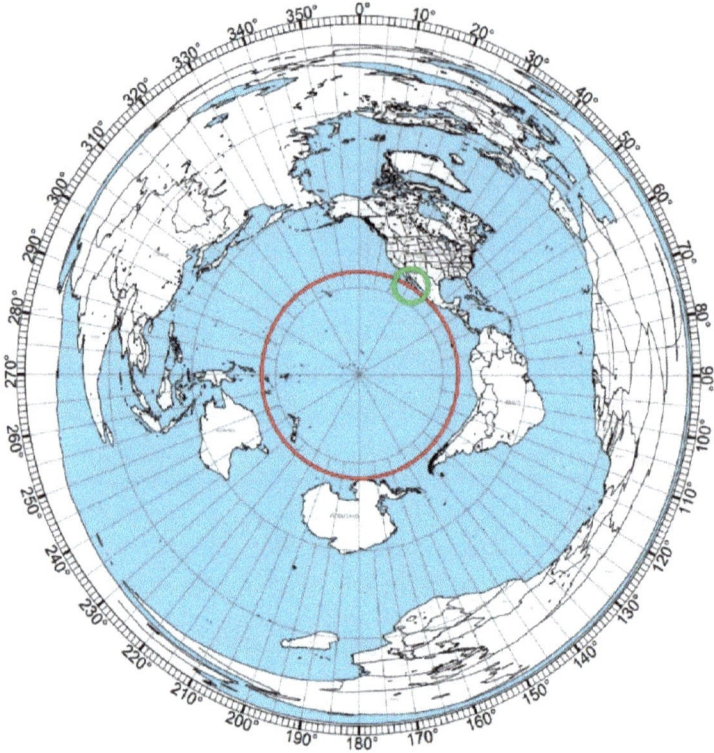

Figure 50–World Azimuthal Map centered in Kaaba's Antipode
Colored circles are added
Courtesy of Tom (NS6T), https://ns6t.net/azimuth/azimuth.html

You can see how the OUTSIDE circular area, close to the center, is mostly water. In contrast, the INSIDE area, which is the outmost, is largely land.

The largest part of Northern America, almost all Southern America

(except eastern Brazil), Antarctica, and Australia are in the MIDDLE2 circular area. Same as the previous Figure. There is one note to consider here, and that is Islam. For Muslims, the **Qiblah** is especially important, and it is the direction to pray and for burial globally. One question that you might have in your mind is the shift of Qiblah in Islam. In Part 2 of this article, I represented that the Qiblah of Islam was the *Al-Aqsa Mosque* in the ***Old City of Jerusalem*** first, from 610 CE. The prophet of Islam changed the direction of Qiblah towards Kaaba in 624 CE. You can refer to some of the Quran verses such as Surah Al-Baqarah, 2:144, and 2:143. Besides religious purposes, perhaps the Qiblah shift has something to do with shifting the lands on Earth.

Thus, we saw how the Kaaba and its antinode point of view split the globe into two perspectives:

- Mostly Land
- Mostly Water

But what about the people who lived before Islam? I showed how the diameters of some Pyramids and the Colossi of Memnon statues are towards Kaaba. They are from before Islam. Maybe the ancient designers somehow knew some important things regarding Giza Pyramids and Kaaba areas unclear yet.

Remember that this book is solely scientific research for knowledge and education. This study is more related to Earth Science, a branch of Geography, and I refer more research to the specialists in this field.

As you can understand, more questions arise each time we study deeper.

In the next section, I will study this matter from another perspective seeking any hidden clues.

Geography and Earth Science

Earth Science is one of the four historical traditions in geographical research. The first person to use the word γεωγραφία (Geography) was *Eratosthenes* (276–194 BCE), born in *Cyrene*, North Africa (now *Shahhat, Libya*). He was a Greek polymath: mathematician, geographer, poet, astronomer and music theorist. He was a man of learning, becoming the chief librarian at the *Library of Alexandria*. He invented the discipline of geography, including the terminology used today.[31][32]

The oldest known world maps date back to ancient *Babylon* from the 9th century BCE.[33]

The *Austronesian* people developed the first sea-going sailing ships from what is now *Taiwan*. Their invention of *catamarans, outriggers*, and *crab claw sails* enabled their boats to sail for vast distances in the open ocean. It led to the *Austronesian Expansion* at around 3000 to 1500 BC.[34]

How did our older ancestors who lived before that time knew about the Land-Water distribution on Earth's surface? What else did they know that we have not discovered yet? There are many questions that we have.

In the next section, I am going to go further back in time.

31. Eratosthenes (20100124). Eratosthenes' Geography. Translated by Roller, Duane W. Princeton University Press. ISBN 9780691142678. Archived from the original on 20151119. Retrieved 20150626.
32. Pattison, William D. (Summer 1990). "The Four Traditions of Geography" (PDF). Journal of Geography (published 1964). September/October 1990 (5): 202–206. doi:
33. Kurt A. Raaflaub & Richard J.A. Talbert (2009). Geography and Ethnography: Perceptions of the World in PreModern Societies. John Wiley & Sons. P.147. ISBN9781405191463
34. Meacham, Steve (11 December 2008). "Austronesians were first to sail the seas". The Sydney Morning Herald. Retrieved 28 April 2019

Part 18
Opposite Land-Water

This section is breathtaking. It seems like the land and water areas are opposite of each other due to the center point of Kaaba.

The Next Figure highlights this matter. Look at the *MIDDLE-2* area where North American, South American, Australia, and Antarctica are set.

1. You can see how, estimate, the whole Australian dryland (red circle) can be reversed (by yellow lines) by the Kaaba spot (brown circle) and fit between the North and South American drylands. Note how the curve of the eastern Australia border matches with Mexico and the Central American countries. Also, the entire New Zealand island fits in the water part of the western shore of Mexico. By tracing the land-borders, you can easily see the matching areas. You can omit the small islands for a better understanding of the concept.

You can see how the other lands also approximately fit into the watery areas across the Kaaba in the same method.

Azimuthal Map

Centered at Kaaba

Center: 21°25'20"N 39°49'34"E

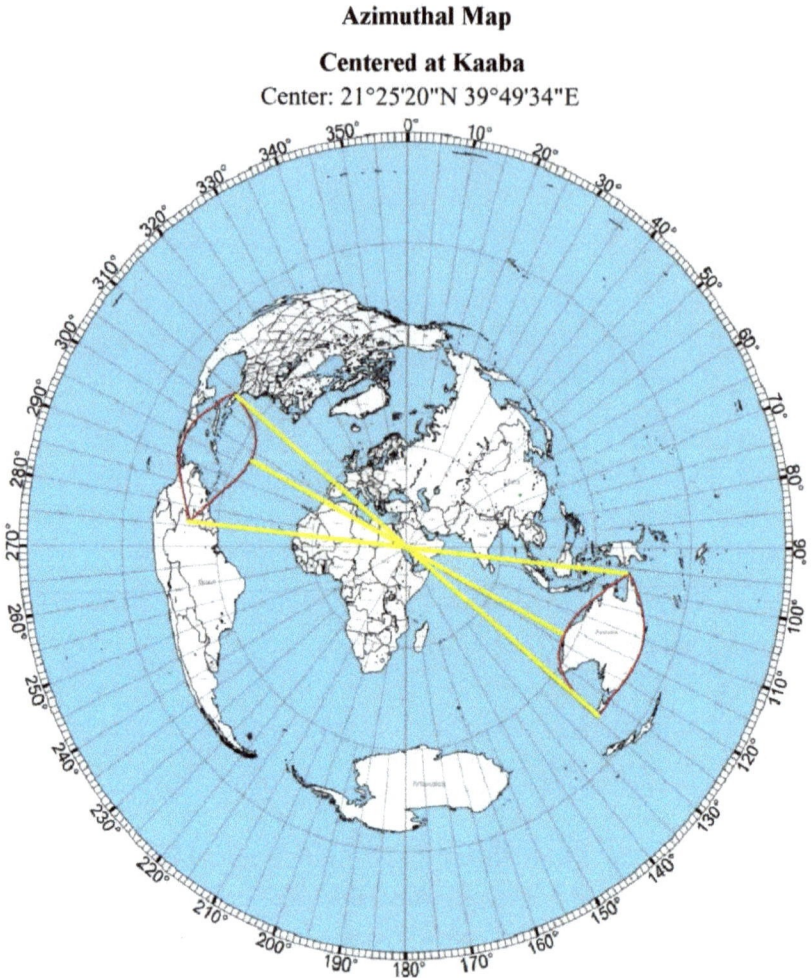

Figure 51 Azimuthal map of the world centered at Kaaba
Reversing the Australian land
Colored arcs and lines are added
Courtesy of Tom (NS6T), https://ns6t.net/azimuth/azimuth.html

2. North America, from Alaska state to Mexico, almost fits between Australia and Antarctica. However, Northern a section of Canada and Alaska State overlap with Antarctica, which also has cold climates.

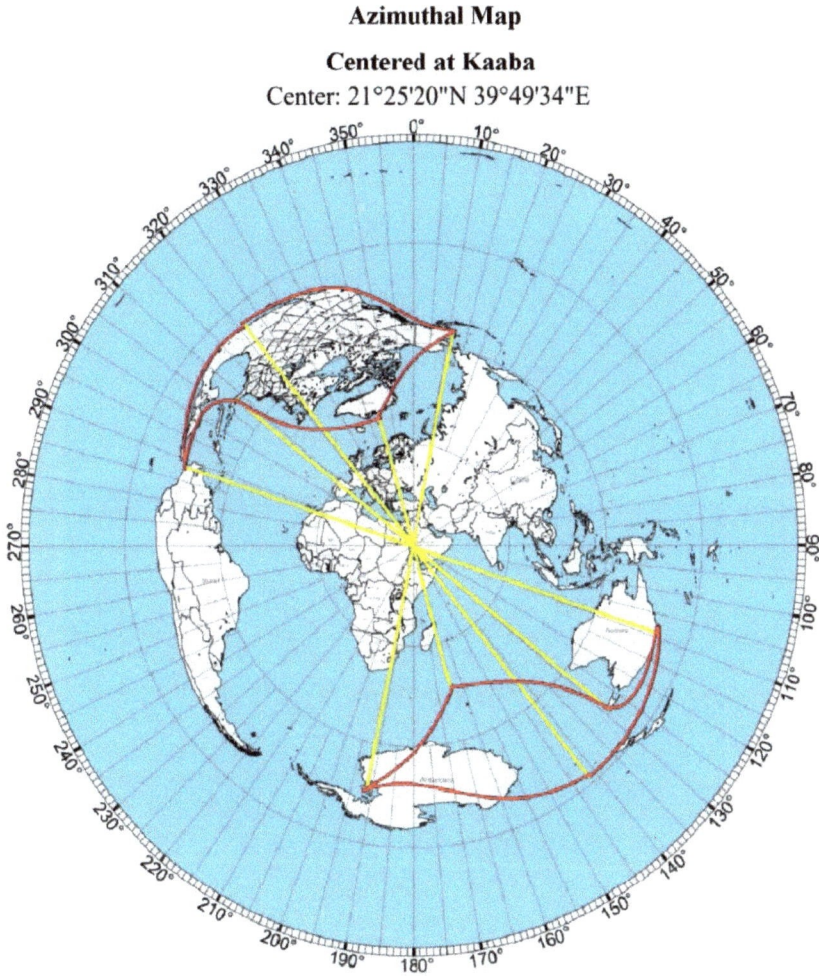

Figure 52 Azimuthal map of the world centered at Kaaba
Reversing the North American land
Colored arcs and lines are added
Courtesy of Tom (NS6T), https://ns6t.net/azimuth/azimuth.html

3. The entire Southern American land fits between Australia and Northern America lands. The watery area between the southern America and Antarctica continents suggests a missing land in the opposite direction.

Azimuthal Map

Centered at Kaaba

Center: 21°25'20"N 39°49'34"E

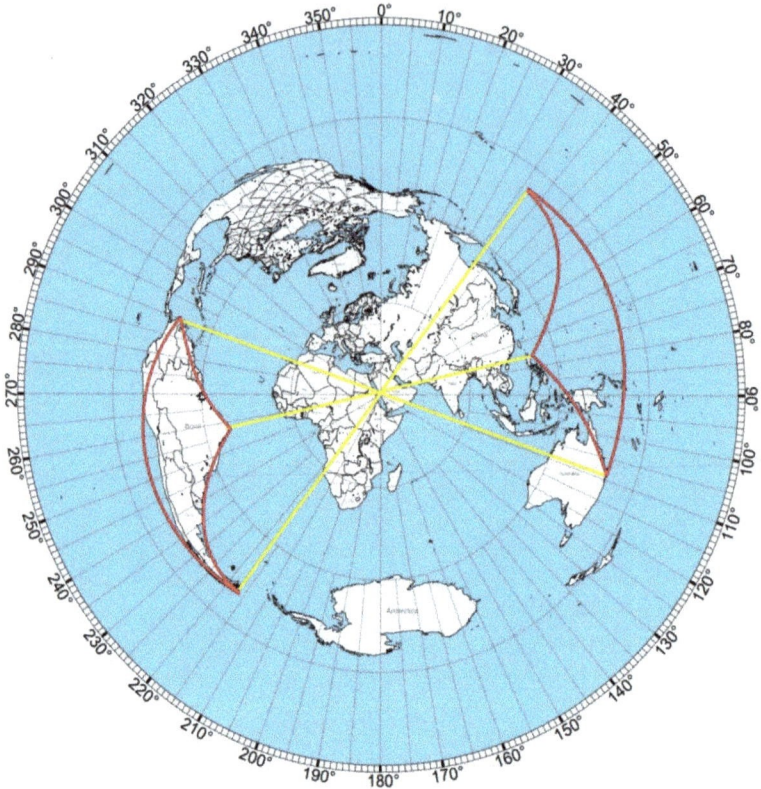

Figure 53–Azimuthal map of the world centered at Kaaba
Reversing the South American land
Colored arcs and lines are added
Courtesy of Tom (NS6T), https://ns6t.net/azimuth/azimuth.html

Thus, this part shows how the land and water are reversed due to the Kaaba spot. I named the missing land as '***Continent M***'. Also, I chose Kaaba as an example. However, more precise calculations are required to seek the exact spot. Anyways, somewhere between Kaaba, Giza Pyramids, and ***Sha'Z Square*** should be the spot that makes the reversing of landwater more geometric.

Part 19
Sha'Z Small Land Circle

Now, let me change the Kaaba centric approach to another one. The next Figure is from exactly above *Taq-e Bostan* at the northern corner of the red *Sha'Z Square*. The yellow circle, which is on Earth's edge from this point of view, is centered at Taq-e Bostan. Radius is 5,297.6 miles. I named this circle as "*Sha'Z Small Land Circle*" for ease.

From Taq-e Bostan to the furthest land spot, the green line is heading 16.37° towards North to Spot TN 64.402371°, -172.233639°. The yellow line is from Taq-e Bostan towards the most distant land Spot TS in South at -34.358161°, 18.475601° heading 204.34° towards the North. The green and yellow lines are almost opposite each other (7.97° difference). They amazingly almost pass over the Sha'Z Square diameters.

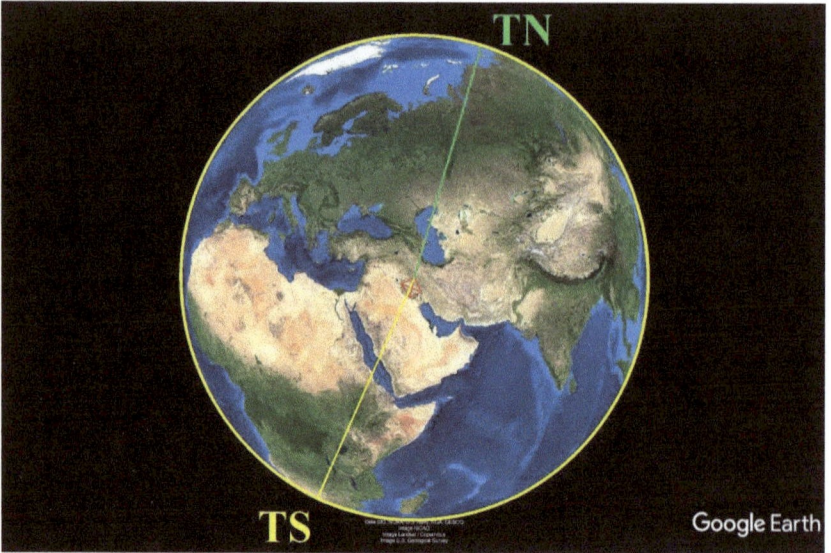

Figure 54–Sha'Z Small Land Circle centered at Taqe Bostan
Yellow and green lines and the circle are added.
Google Earth Figure date: 12/13/2015, eye alt: 8822.02 mi
IMAGE: Landsat / Copernicus

Azimuthal Map
Centered at Taq-e Bostan
Center: 34°23'15"N 47°7'55"E

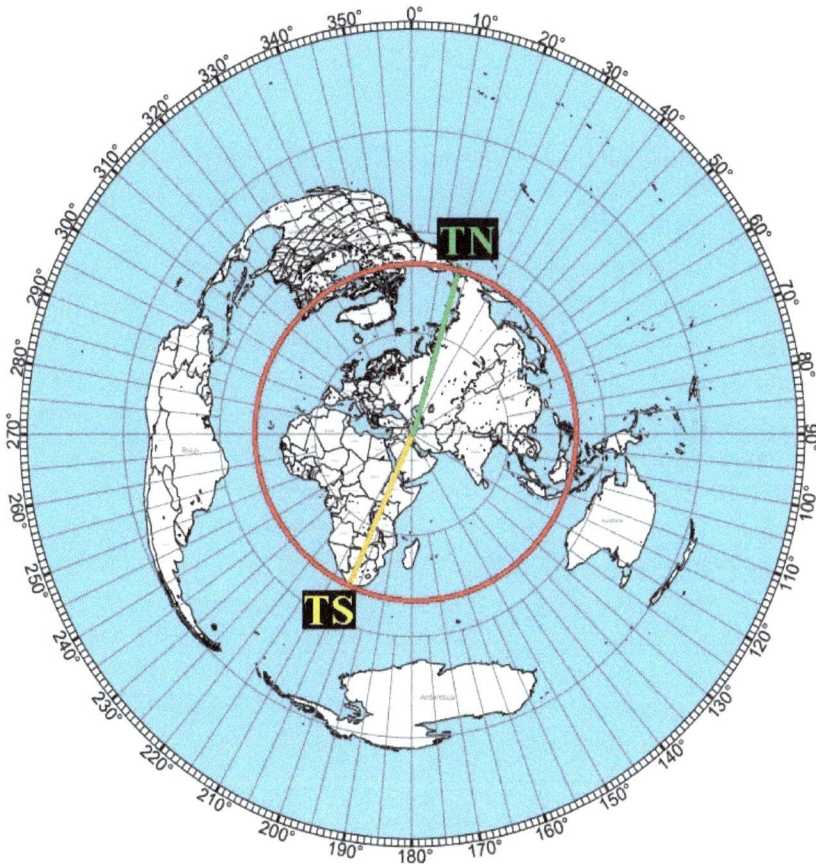

Figure 55–Azimuthal map of Sha'Z Small Land Circle
Red circle, lines, and letters are added
Courtesy of Tom (NS6T), https://ns6t.net/azimuth/azimuth.html

It is fascinating how geometrically, *Sha'Z Small Land Circle* almost surrounds Earth's edge from about 9,000 miles above the ground. I do not know if our ancestors who lived on the surface of Earth several thousand years ago knew anything about this or not. What if they knew?

The main drylands of Africa, Europe, North Pole and Asia, except some parts of Southeastern islands, are surrounded by the *Sha'Z Small*

Land Circle (refer to Figures 45 and 46). Excitingly, this Red Circle is almost on the Earth's edge (refer to Figures 46). Is this coincidence?

The most distant land from Taq-e Bostan would be the Spot TN near the top of the Figure (at the end of the green line) at 64.401432°, -172.232369°. It is 5,297.62 miles away and heading 16.37° from the North. It's rounded latitude to the tenth place is 64.4°, in Group 2.

Amazingly, from the other side, the most distant land from Taq-e Bostan towards Africa continent would be the Spot TS (at the end of the yellow line near the bottom of the Figure) at -34.357497°, 18.473100°. It is 5,080.05 miles away and heading 204.35° from the North. Its rounded latitude to the tenth place is 34.4°, also in Group 2.

Spot	Distance from Taq-e Bostan (miles)	Angle from Taq-e Bostan (Clockwise)	Group
Spot TN	5,297.62	16.37°	2
Spot TS	5,080.05	204.35°	2
	217.57 (%0.021)	187.98° (7.98° more than 180°)	

Table 17 Spot TN and Spot TS comparison

Their distances to *Taq-e Bostan* are alike, with only about 217.57 miles (350.15 Kilometers) difference in about 10377.67 miles (%0.021).

By paying attention to the digits of their rounded latitudes to the tenth place, you can see they both have four. And they are almost oposite of each other with the approximately same distance. You might ask, are the numbers of these two spots coincidence? Is Taq-e Bostan a random place?

Rhombus shape

Another exciting geometric relation is the rhombus shape of the Mainland.

The next Figure represents this matter. The Red Circle is "**Sha'Z Small Land Circle,**" centered at **Taq-e Bostan**.

Spot TE is the most Eastern location, and Spot TW is the most western location from Taq-e Bostan on the Mainland.

Spot TW is at 12.404928°, -16.771425°, in Group 2. About 4,259.11 miles (6,857.37 Kilometers) from Taq-e Bostan with the headings of 265.74° (clockwise) from the North. This angle is 94.26° counterclockwise from the North, which is only 4.26° more than 90° and almost a right angle.

From the other side, Spot TE is 28.285238°, 121.638391° Its rounded latitude to the tenth place is 28.3°, which puts it in Group 2. Note that the latitude needed only 0.06° more to become rounded to 28.4° in Group 2, which is omittable. Spot TE is almost 4,326.79 miles (6,963.29 kilometers) away, with the headings of 73.08° (clockwise) from the North. This number is only 3.08° more than 70° and about 17° less than a right angle with 90°.

Spot	Distance from Taq-e Bostan (miles)	Angle from Taq-e Bostan (Clockwise)	Group
Spot TE	4.326.79	73.08°	2
Spot TW	4,259.11	265.74°	2
	67.68 (%0.008)	192.66° (12.66° more than 180°)	

Table 18 Spot TE and Spot TW comparison

Both of the distances to *Taq-e Bostan* are remarkably alike, with only about 67.68 miles (105.92 Kilometers) difference in about 8,586 miles (%0.008).

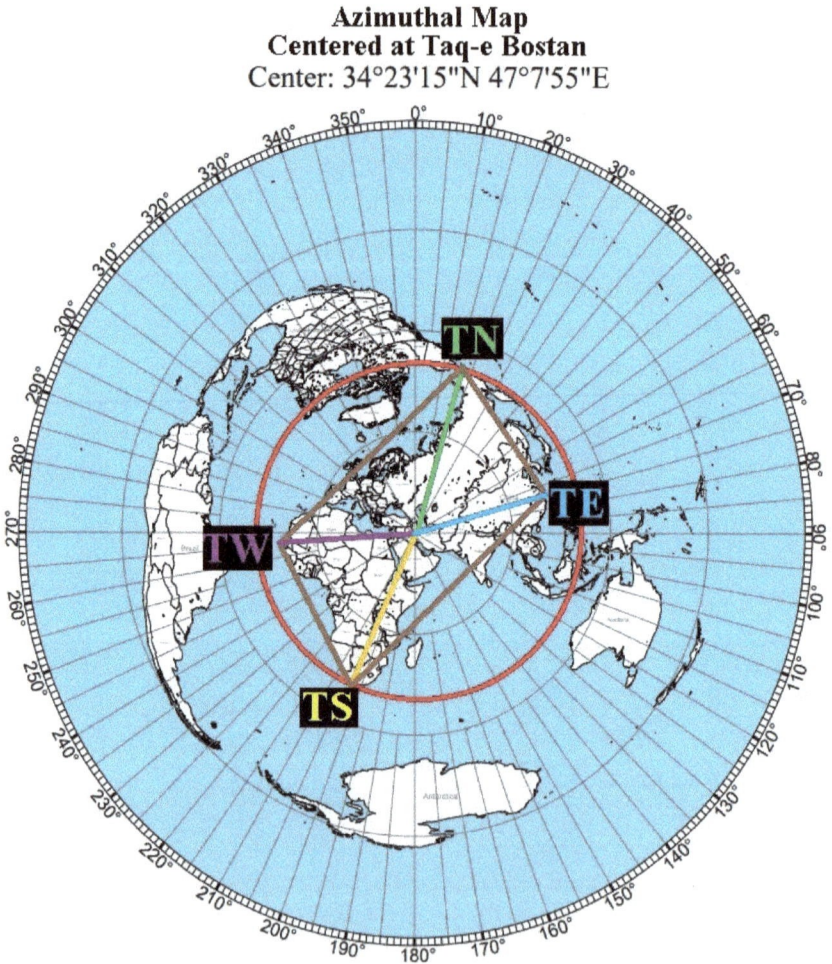

Azimuthal Map
Centered at Taq-e Bostan
Center: 34°23'15"N 47°7'55"E

Figure 56–Azimuthal map of Sha'Z Small Land Circle including Mainland's Rhombus Shape
Colored circle, lines, and letters are added
Courtesy of Tom (NS6T), https://ns6t.net/azimuth/azimuth.html

Some exciting mathematical relations can be seen in between these spots. For example, all four corners are in Group 2, and the diameters are almost straight lines, nearly 180°. Is this accidental?

If I want the diameters to be perfectly straight lines, I should slightly alter the Red Circle center from Taq-e Bostan to somewhere around Syria on the flat Azimuthal map. But on the Globe is different. The intersection is Spot TC at 42.904077°, 30.771018° at the Black Sea's western side by Google Earth features. Rounded Latitude puts this Spot in Group 1.

Also, TN and TE's angle at the center of Taq-e Bostan is 56.71°, only 3.29° less than 60°. On the other side, TS and TW's angle is 61.40°, only 1.40° more than 60°. So, the angles are extremely similar to 60°. Is this another Coincidence?

In addition, the brown Rhombus Shape of the Mainland is almost symmetric. Look at the land borders. You can see how the curvy indentborder from Spot TW to Spot TN is similar to the curvy indentborder from South Africa to Eastern China. Also, the curvy indent-border from TW to TS is like the outdent-border of TE to TN.

Land Separation

Looking at the Azimuthal Map above, you might produce some other fascinating geometrical relations. Here, I'm going to show you some exciting erudition and outcomes.

In part 18, Opposite Land-Water, I showed how the Lands are set. Now, I want to show how the diameters and the Rhombus Shape medians almost separate those lands.

The next Figure displays how the medians and diameters' extended lines from the Rhombus Shape separate North America, South America, Australia and Antarctica lands.

The Pink Lines are the median lines from the brown Rhombus Shape.

I added the letters, and approximate lines and shapes to understand the context easier. They are not accurate.

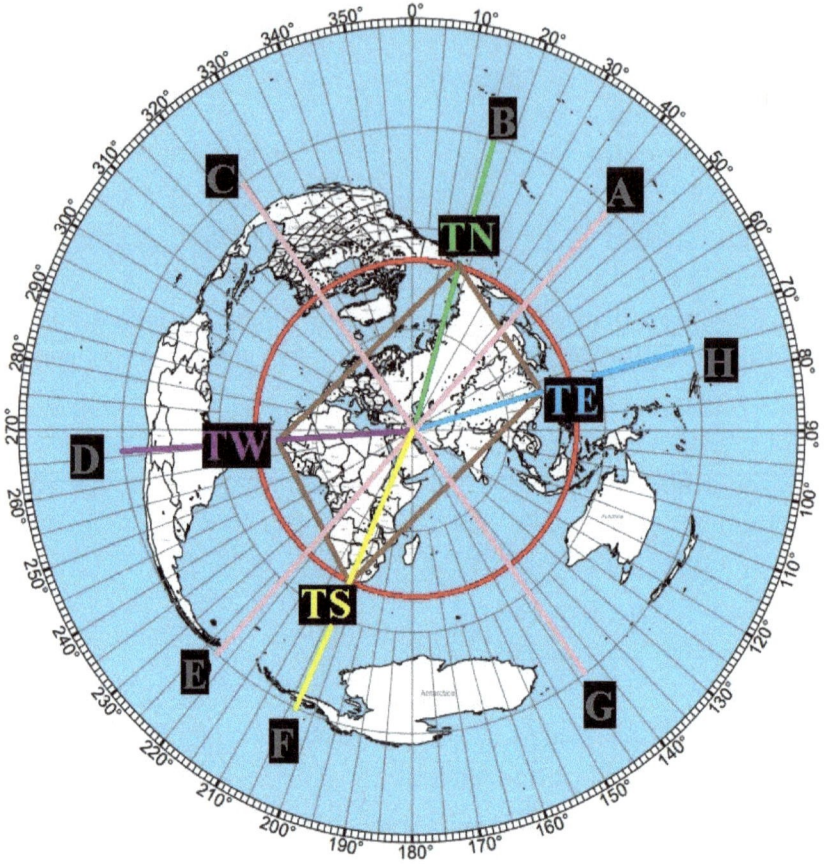

Figure 57–Azimuthal map of Sha'Z Small Land Circle including Mainland's Rhombus Shape with land separation
Colored circle, lines, and letters are added Courtesy of Tom (NS6T), https://ns6t. net/azimuth/azimuth.html

- Lines B and C are almost separate North America.
- Line D crosses almost through the middle of South America.
- Lines C and E almost separate Central and South America.
- Lines E/F and G almost separate Antarctica.
- Lines G and H separate Australia.

Perhaps the ancient Persians, who built the Taq-e Bostan site, had access to all of this knowledge in their era to choose the location for their structures. And you might want to understand some correlation between water, land, Sun, and religion in this matter.

For further study, you can use other centric locations, such as the Giza Pyramids, to see if you can develop the Mainland Center's exact location and produce fascinating information.

Part 20
Water in Body

Building many of the structures were not in the same era or even area. Yet, they pose as if there were full awareness of each other's precise location, which studying them could radically change our understanding of human history. There are many clues that our ancestors showed interest in water, land, the Sun and religion, yet many questions are still hovering for answers, which might be right in front of our eyes.

For this section, I will continue the study by examining inside our body, physically and intellectually, to find more clues.

Each day, people must consume a certain amount of water to survive. In physiology, body water is the water content of an animal body in the tissues, the blood, the bones and elsewhere. Water is of major concern to all living things; in some organisms, up to 90 percent of their body weight comes from water. According to *H.H. Mitchell*, Journal of *Biological Chemistry* 158, the brain and heart are 73 percent water, and the lungs are about 83 percent water. The skin contains 64 percent water, muscles and kidneys are 79 percent, and even the bones are watery: 31 percent. According to *Dr. Jeffrey Utz*, Neuroscience, Pediatrics, *Allegheny University*, different people have different percentages of their bodies made up of water. Babies have the most, being born at about 78 percent. In adult men, about 60 percent of their bodies is water. In adult women, this number is near 55 percent.[35]

Like our Earth's surface with about 70 percent water, most of our body holds water. This number is a commonality between the human race and the surface he lives on. Per our daily dehydration, we need to consume water constantly.

Yet, we do not know the origin of how humans learned to consume water and how they knew water evaporates from their bodies due to heat. Perhaps they understood it is hotter when the Sun is directly above them, and that is why they get thirstier that way. Thus, there might be a clue in the Sunlight.

---

35. https://www.usgs.gov/specialtopic/waterscienceschool/science/wateryouwaterand humanbody?qt-science_center_objects=0#qtscience_center_objects

Part 21
Sunlight

Sun is the brightest object in the sky. If our ancestors were interested in the center of drylands, maybe the fact is hidden within the rays of Sunlight, perhaps its temperature. I mean, Sunlight has heat, and the amount of solar reflection and absorption has a profound influence on Earth's temperatures that indicates warming.

One fact of ***Sunlight Absorption*** is color. Darker colors absorb more heat energy. Water on Earth's surface is darker than land; thus, it absorbs more Sunlight energy. White has the least absorption, and our lower classes' Egyptian ancestors wore dresses made of white or unbleached fabric. Since the earth's water is darker and less shiny than its land, water should have more Sunlight absorption.

Imagine a scenario when Sun is directly above the Kaaba's Antipode; in this case, Earth's surface absorbs the maximum Sunlight energy almost only by its waters due to the highest amount of water distribution from the Sun's perspective. Refer to Figures 47 and 50.

Sha'Z Square

And imagine another scenario when Sun is directly above the Kaaba itself; in this case, Earth's land has its ultimate chance to absorb Sunlight versus water. Refer to Figures 46 and 49.

Interestingly, the whole Earth's surface has about 70 percent water and 30 percent land been maybe to keep the balance of Sunlight absorption.

Angular Distribution

Another fact is the **Angular Distribution** of Sunlight reaching the surface. At noon, the body exactly underneath Sun absorbs most heat. Meanwhile, the Sunrise and Sunset areas absorb the least.

I want to see what Muslims do in that realm. I chose Muslims not to promote them but because they are the world's secondlargest and fastest-growing religion that principally resides in the area between Kaaba, Giza Pyramids, and **Sha'Z Square**.[36,37]

Sunlight's reflection and absorption may strongly correlate with the Muslim prayer's timing within the Sunrise, Noon, and Sunset. Also, Muslims require pure water and soil, besides the direction to Kaaba for their prayers.

Besides, Muslims wear white clothing during the **Ihram pilgrimage** (**Hajj**) and or (**Umrah**), which gets the least Sunlight absorption. Mans Ihram differs from women, and while man's Ihram is two sheets of white cloths that leave some parts of his body naked.

You might want to know if there are any correlations that Muslims cover Kaaba with black fabric, and they placed the **Black Stone** on it. The Black Stone (Arabic: **al-Hajaru al-Aswad**) is a rock set into the Eastern corner of the Kaaba. According to Islamic tradition, it is revered by Muslims as an Islamic relic, dates back to **Adam** and **Eve's** time.

People worship the Sun in ancient times. Although almost every ancient civilization used solar motifs, only relatively few cultures (Egyptian, Indo-European and Meso-American) developed solar religions.

---

36. Sherwood, Harriet (27 August 2018). "Religion: why faith is becoming more and more popular". The Guardian. ISSN 0261-3077. Retrieved 13 January 2021.
37. "World Population Clock: 7.8 Billion People (2021) Worldometer".
www.worldometers.info. Retrieved 13 January 2021.

Solar deities, gods personifying the Sun, are sovereign and allseeing. In ancient Egypt, the Sun-god was the dominant figure among the ultimate gods and retained this position early in that civilization's history. Sun heroes and Sun kings also occupy a leading role in Indian mythology.

During the later periods of Roman history, sun worship gained importance and ultimately led to "*Solar Monotheism*."[38,39,40,41,42,43,43]

38. Sheikh SafiurRehman alMubarkpuri (2002). ArRaheeq AlMakhtum (The Sealed Nectar): Biography of the Prophet. DarusSalam Publications. ISBN 9781591440710.

39. HendersonSellers, A.; Wilson, M. F. (1983). "The Study of the Ocean and the Land Surface from Satellites". Philosophical Transactions of the Royal Society of London A. 309 (1508): 285–29 4. Bibcode: 1983RSPTA.309.285H. doi:10.1098/rsta.1983.0042.

40. Coakley, J. A. (2003). J. R. Holton and J. A. Curry (eds.). "Reflectance and albedo, surface" (PDF). Encyclopedia of the Atmosphere. Academic Press. pp. 1914–1923.

41. Thompson Gale (2003). Environmental Encyclopedia (3rd ed.). ISBN 9780787654863.

42. Cornell, Vincent J. (2007). Voices of Islam: Voices of tradition. Greenwood Publishing Group. p. 29. ISBN 9780275987336. Retrieved 26 August 2012.

43. Leventon, Melissa (2008). What people wore when: A completely illustrated history of costume

Part 22
Prayer

According to **Britannica, Adalbert G. Hamman,** "Prayer, an act of communication by humans with the sacred or holy God, the gods, the transcendent realm, or supernatural powers. Found in all religions in all times, prayer may be a corporate or personal act utilizing various forms and techniques."[44]

William James and psychologists such as **Joseph Segond** describe Prayer as a "**Subconscious**" and "**Emotional Effusion**," an outburst of the mind that desires to enter into communication with the invisible.[45]

Prayer forms in the world's religions, though varied, generally follow certain fixed patterns. From the third millennium BCE to the beginning of the Common Era, Prayer forms changed little among the Assyrians and Babylonians and their descendants.[46]

There are preserved Ancient Egyptian piety in numerous precepts engraved on the backs of scarabs. These engravings sometimes include praises of the divinity ("All good fates are in the hand of God"), statements of confidence, or requests for protection for the one praying and for his whole family ("God is the protector of my life; the house of one favored by God fears nothing").[47]

44. PRAYER IN IGBO COSMOLOGY: THE CASE OF MBIERI, MBAITOLI L https://www.ajol.info/index.php/ijdmr/article/download/79304/69606
45. Prayer | Britannica. https://www.britannica.com/topic/prayer
46. Prayer Forms of prayer in the religions of the world ....
https://www.britannica.com/topic/prayer/Formsofpray erinthereligionsoftheworld
47. Prayer Forms of prayer in the religions of the world ....

Judaism is one of the best-known collections of prayers, the 150 psalms in the Bible. In these psalms, which always presuppose a collective witness, though they may be used by an individual privately, praise is descriptive (God is…) or narrative (God does…) in nature.[48]

In Chinese Buddhism and Daoism, in addition to Prayer that accompanies sacrifice, there is the **monastic Prayer** (*muyou*), which they practice morning, noon, and night to the sound of a small bell. [49]

From its beginning in the seventh century CE, the most important part of Islamic liturgy has been the ritual prayer called the al t (daily Prayer), in which Christian and Jewish influences can be seen. This minutely detailed Prayer is while the supplicant face Kaaba (in Saudi Arabia) five times a day. These prayers may practice alone if one is unable to go to the mosque. The mosque, "**Masjid**" in Arabic, is the Muslim gathering place for Prayer.

Masjid means "**Place of Prostration**." The home of the Prophet Mohammad is the first mosque. [50]

The style, layout and decoration of a mosque can tell us a lot about Islam in general and its construction era and territory. The varied architecture of mosques is usually by the regional traditions of the time and region. In the Islamic areas, a mosque in its many forms is the quintessential Islamic building. Certain architectural features appear in most of them, such as the "**Mihrab**," a niche in the wall that indicates the Qiblah (direction of Kaaba). Most mosques also feature one or more domes, called "**Qubba**" in Arabic, a symbolic representation of Heaven's vault.[51]

Per the discussion in Part 7 about the shape of pyramids, Qubba, usually at the mosques' top, does not have side edges like the SquareBased Pyramids and Cones are not like igloo structures half-sphere. They are

https://www.britannica.com/topic/prayer/Formsofprayerinthereligionsoftheworld
48. Prayer Religions of the East | Britannica.
https://www.britannica.com/topic/prayer/ReligionsoftheEast
49. Prayer Religions of the East | Britannica.
https://www.britannica.com/topic/prayer/ReligionsoftheEast
50. Prayer Religions of the East | Britannica.
https://www.britannica.com/topic/prayer/ReligionsoftheEast
51. Introduction to mosque architecture – Smarthistory.
https://smarthistory.org/introductiontomosquearchitecture/

still wide in the lower part and narrow at the top, similar to the Japanese *Suehirogari* shape–/\–and the Persian *Hasht* shape —, which both mean the number eight.

Regardless of the religious perspectives, one might think that mosques are more advanced spiritual structures because they are stable and have a Mihrab that indicated the direction to Kaaba. Mosques also have Qubba at the top to express the rise to Heaven besides the more immeasurable resistance against edge-erosions. Where did the Islamic architects depict these designs for their Mosques?

Muslims perform their first Prayer before sunrise, the second at noon, the third in the late afternoon, the fourth immediately after sunset, and the fifth late evening before retiring to bed. This timing means that the Sun plays a significant role in Prayer. Islamic Prayer is an act of adoration of Allah (God), and thus it would not be suitable to add a request. Before adoring God, the believer must purify himself utilizing ablutions in pure water or, failing this, in the sand. [52,53]

You might ask why the Sun is instead of a planet or the Moon. Note that the cubic structure of Kaaba has an important shape difference with Pyramids, which is the roof. Pyramids do not have tops, representing a Kaaba philosophy that assigns number *1* to the Allah/ God due to its solitary.

Thus, by studying deeper, many questions arise. Anyway, the Sun's importance is beyond just the brightest object in the sky for our ancestors..

By adding everything together, there is an important correlation between the Sun and the Mainland Center, near *Sha'Z Square*, for our ancestors. And that is why some of our ancestors built many structures to sign towards the Mainland Center directly.

---

52. https://www.britannica.com/topic/prayer
53. https://www.britannica.com/topic/prayer/Formsofprayerinthereligionsoftheworld #ref66259

Part 23
Summary

Whenever we discover something new, we find surprises. This study opens the mind to look at ancient sites from different perspectives and drags our minds to contemplate deeply. There are some footprints left by our ancestors that represent an interest in some locations such as Kaaba, Giza Pyramids, and *Sha'Z Square* places. What they illustrated is still baffling us.

At first glance, this research compounds some ancient locations to each other with deep reverence to arts, geoscience, mathematics, and worships.

However, the reason our ancestors were interested in these locations, the shape of structures, and the message they dispatched remain a mystery. It is lucid that our ancient designers captivated mathematics and geometry in building their magnificent buildings at such early periods. But it is a wonderment that our ancestors had access to some extensive knowledge that made them aware of these particular spots on Earth's surface.

Some parts of this research might even burst into ideas that contradict a normal understanding of our ancient civilization's evolution involving their mathematicians and designers. As an example, one obvious difference between the two views of Earth, from Kaaba, and its antipode, is the portion of water-land on the Earth's surface. It seems like somewhere near Kaaba has a central identity of all the main drylands on Earth's surface.

It also seems like somewhere between Kaaba, Giza Pyramids and **Sha'Z Square** is almost in the Mainlands center between Africa, Europe and Asia. And finally, it seems like the Sun's position when it is right above the Mainlands center was a concern for our ancestors.

Overall, although these ancient mysteries may never become fully revealed, the enigmatic items we do not yet understand might not be random. We need to come out of the darkness and identify forgotten episodes of our ancestor's astounding knowledge and concerns.

Perhaps, additional clues still lie hidden, etched on our ancient structures and locations, right before our eyes, and waiting to be unveiled. Further investigation and a love for our universe and the old world will reveal them.

Part 24
Supplements

The next table displays some information regarding the entire historical sites in this book:

Group	Name	Approx. Built time	Latitude	Longitude (degrees)	Latitude (rounded)
1	Arc de Triomphe du Carrousel	1806 Started	48.861750	2.332887	48.9
1	Arc Du Triomphe de l'Étoile	1806 Started	48.873785	2.295152	48.9
1	Temple of Apollo, Patroos	340-320 BCE	37.975542	23.722099	38.0
1	Ziggurat Chogha Zanbil	1,250 BCE	32.008333	48.520833	32.0
1	Ziggurat of Ur	21st century BCE	30.962778	46.103056	31.0
1	Menkaure	2,580-2,560	29.972497	31.128278	30.0

1	Sphinx	unknown	29.975263	31.137558	30.0
1	Khafre	2,580-2,560	29.975997	31.130733	30.0
1	Khufu	2,580-2,560	29.979147	31.134219	30.0
1	Kaaba Zartosht	550–330 BCE	29.988889	52.874722	30.0
1	remains of Persepolis	515 BCE	29.934444	52.891389	29.9
1	Church of Saint George	late 12th or early 13th century AD	12.031625	39.041145	12.0
1	La Huaca	4,000 - 4,600BCE	-10.893566	-77.520412	-10.9
1	The Gallery	4,000 - 4,600BCE	-10.893497	-77.519401	-10.9
1	Pyramid of Lesser	4,000 - 4,600BCE	-10.891820	-77.520292	-10.9
1	Christ the Redeemer statue	between 1922and 1931	-22.951944	-43.210556	-23.0
2	Spot TW	Unknown	12.404928	-16.771425	12.4
2	Spot TE	Unknown	28.285238	121.638391	28.3
2	Spot TN	Unknown	64.402371	-172.233639	64.4
2	Spot TS	Unknown	-34.358161	18.475601	-34.4
2	Tumulus of Bougon	4,800 BCE	-46.373795	0.066660	46.4
2	The site of Taq-e Bostan	4th century AD	34.387529	47.132096	34.4

2	Ziggurat of Dur-Kurigalzu	14th century BCE	33.353611	44.202222	33.4
2	Atlantic Anomaly Rectangle	Unknown	31.400000	-24.000000	31.4
2	Kaaba	Unknown	21.422500	39.826184	31.4
2	Kaaba Antipode	Unknown	-21.408745	-140.194917	-21.4
3	Leshan Giant Buddha	between 713 and 803	29.544722	103.773333	29.5
4	Tumulus of St.Michel	4,500 BCE	47.588244	-3.073689	47.6
5	Colossi of Memnon	1,350 BCE	25.720469	32.610476	25.7

Table 19 Historical sites in this article rounded latitudes

Other Sites

Many historical sites are available for further study. Thanks to modern technology that makes the investigations easier these days. Some ancient sites are arranged below for curious and determined scholars to study these topics moreover.

Los Millares is a 4.9 acres area in Santa Fe de Mondújar, Andalucía, Spain.
It is at 36.964722°, -2.522222° that was active in the fourth millennium to the second millennium BC. This site's rounded latitude in the tenth place is 37°, which puts it in Group 1. Interestingly, the angle to Kaaba is exactly 101° from the North.[54]

---

54. https://en.wikipedia.org/wiki/Los_Millares

Kot Diji was the forerunner of the Indus Civilization. The occupation of this site is attested already at 3,300 BCE, at 27.345556°, 68.706667°. This site's rounded latitude in the tenth place is 27.4°, which puts it in Group 2.[55]

Nausharo is in Balochistan, Pakistan, and well known as an archaeological site for the Harappan period. It was between 3,000 and 2,550 BCE. At 29.365°, 67.588°. This site's rounded latitude in the tenth place is 29.4°, which puts it in Group 2.[56]

Mehrgarh is a Neolithic site on the Kacchi Plain of Balochistan, Pakistan, 7000 BCE. At 29.383333°, 67.616667°. This site's rounded latitude in the tenth place is 29.4°, which puts it in Group 2.[57]

Yonaguni Monument- In the Japanese territory, there is a submerged rock formation off the coast of Yonaguni. At 24.435833°, 123.011389°. Some believe it is not an ancient human-made structure. However, this site's rounded latitude in the tenth place is 24.4°, which puts it in Group 2. I wonder if it is a human-made structure.[58]

Carnac Stones at 47.5965°, -3.066°, which is stretched almost 60° from the North. This site's rounded latitude in the tenth place is 47.6°, which puts it in Group 4. If you draw a line from Taq-e Bostan to this location, surprisingly, the angle would be about 90° from the North.[59]

55. https://en.wikipedia.org/wiki/Kot_Diji
56. https://en.wikipedia.org/wiki/Nausharo
57. https://en.wikipedia.org/wiki/Mehrgarh
58. https://en.wikipedia.org/wiki/Yonaguni_Monument
59. https://en.wikipedia.org/wiki/Carnac_stones

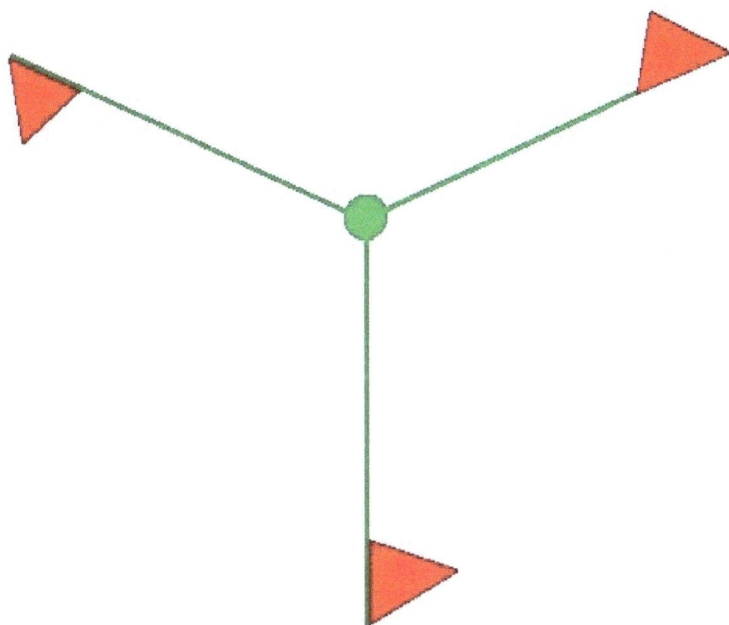

About the Author

Shahrokh Zadeh was born and raised in Iran and has been a resident of the United States of America since 2009. He received his Bachelor of Science in Solid States Physics in 1994 and his Master's in Management Information Systems (MIS) in 2019. Shahrokh has always been fascinated with scientific disciplines. Scrutinizing these topics over the past 27 years, which at times was daunting and overly complex, provided him the confidence to publish his conclusions.

Index

G

Geographic Coordinate System, 1-3, 83
Geography, 1, 2, 113, 114
Geometry, xi, xii, 4, 24, 27, 49, 56, 60, 63, 74, 77, 89, 99, 139
Giza Pyramids, 31, 37-40, 45-50, 54, 55, 63, 64, 74, 76, 82, 90, 91, 94, 95, 98-100, 101, 103, 113, 118, 127, 132, 139, 140
God, xii, 41, 42, 56, 57, 132, 133, 135, 136, 137
Greece, xi, 2, 18, 19

H

Historical dictionary of France, 66 History, xii, 66, 79, 129, 133
History and Practice of Ancient Astronomy, 1

I

Invention of a Geographic Coordinate System, 1
Iran, 2, 15, 29, 33, 34, 35, 37, 147

Iranian, 2
Iranian plateau, 2
Islam, xiii, 23-25, 29, 31, 58, 107, 112, 113, 136,
Islamic, 23, 32, 132, 136, 137,
Isma'il, 24
Isosceles Triangle, 27, 39, 40, 87

K

Kaaba, 32, 33, 34, 35, 47, 63, 83, 115-118; 119, 131, 132, 136, 137, 139, 143
Kaaba Zartosht, 32, 33, 34, 35, 47, 63, 83, 142
Kaaba Zartosht structure, 32
Kaaba's Big Land Circle, 106-111
Kaaba's roof, 26
Kaaba's top, 25
Kermanshah, 2, 29
Khufu, 45-48, 53, 74

L

Land, 105-113
Land-Water, 114, 115, 118, 125
Latitude, 1, 3, 4, 21, 25, 34, 41, 47, 52, 61, 62, 67, 68, 73, 74, 76, 82, 83, 90, 93, 97, 98, 106, 109, 122, 123, 125,
Leshan Giant Buddha, 71-74, 76, 77, 93, 143

Tumulus of St. Michel, 93, 94,
143

V

Verse 127, 24
Verse 96, 23

W

World Azimuthal Map, 110, 112

Z

Zigguratxii, 2, 4, 8, 9, 13, 15,
19, 22, 34,
Ziggurat of Chogha Zanbil, 3, 8,
9, 10, 16, 17, 21
Ziggurat of Dur-Kurigalzu, 2, 3,
5, 6, 8, 10, 17
Ziggurat of U, 2, 3, 6, 7, 8, 17,
142

www.ingramcontent.com/pod-product-compliance
Lightning Source LLC
Chambersburg PA
CBHW040857210326
41597CB00029B/4875